Comets and the Origin of Life

Comets and the Origin of Life

Janaki Wickramasinghe
Chandra Wickramasinghe
William Napier
Cardiff University, UK

World Scientific

NEW JERSEY • LONDON • SINGAPORE • BEIJING • SHANGHAI • HONG KONG • TAIPEI • CHENNAI

Published by

World Scientific Publishing Co. Pte. Ltd.

5 Toh Tuck Link, Singapore 596224

USA office: 27 Warren Street, Suite 401-402, Hackensack, NJ 07601

UK office: 57 Shelton Street, Covent Garden, London WC2H 9HE

British Library Cataloguing-in-Publication Data
A catalogue record for this book is available from the British Library.

COMETS AND THE ORIGIN OF LIFE

ISBN-13 978-981-256-635-5
ISBN-10 981-256-635-X

Printed in Singapore.

Preface

The contemporary scientific approach to the origin of life is being shaped within the emergent discipline of astrobiology which combines the sciences of astronomy and biology. The widespread distribution of water and complex organic molecules in the universe is leading scientists towards a possibly erroneous point of view that life is not only present everywhere but that it is readily generated *in situ* from non-living matter. The idea that water and organics under the right physical conditions lead easily to life has no empirical basis at the present time, nor indeed do we have any definite knowledge of how such a transition occurs. On the other hand, empirical science is now in a position to address the question of whether life can be transferred from one astronomical setting to another.

The search for exosolar planets, life on Mars and elsewhere in the solar system, and dynamical studies of how particulate material can be transferred between potentially habitable cosmic sites, all have a bearing on the question of our origins. We argue in this book that the production of life in the first instance might be an exceedingly rare event but that its subsequent evolution and dispersal are a cosmic inevitability.

The astronomical origin of the 'stuff of life' at the level of atoms is beyond dispute. The chemical elements that make up living systems were unquestionably synthesised from the most common element hydrogen in nuclear reactions that take place in the interiors of stars. Supernova explosions scatter these atoms into interstellar clouds, and new stars and planets form from this material. The combination of atoms into organic molecules can proceed in interstellar clouds via well-attested chemical pathways, but only to a limited level of complexity that falls well short of life. The discovery of biochemical molecules in space material, including in meteorites, arguably crosses this threshold.

In the view of the authors of this book, the interpretation of interstellar organic molecules as the combined product of abiotic synthesis and biological detritus is an emerging paradigm. Inorganic processes can scarcely be expected to compete with biology in the ability to synthesise biochemicals, and if biology is readily distributed on an astronomical scale, its detritus must contribute to the stuff between the stars.

The Aristotelian notion that life could arise readily from everyday materials — fireflies from morning dew — came to be known as the doctrine of spontaneous generation and this doctrine dominated science well into the 19th century. When Louis Pasteur challenged this ancient idea by showing that microbes always arose from pre-existing microbes, the case for panspermia emerged. For if life always derives from pre-existing life, then the possibility must be considered that it predates the Earth. This was the chain of logic followed by Lord Kelvin amongst others in the closing decades of the 19th century.

When Fred Hoyle and one of the present authors re-examined such arguments in the 1970s we turned to comets as the most likely astronomical objects that were relevant to panspermia. In the past three decades considerable progress has been made in geochemistry, microbiology and cometary studies, all of which place comets in the forefront of studies relevant to the origin of life. The basic structure of the present book started as the PhD thesis of the principal author with additional reviews and discussions that bring the whole story up-to-date. Several astrobiology texts have been published over the past decades, but they have been woefully short in their treatment of cometary panspermia. The present book is intended to fill this gap.

We are grateful to the Astrobiology Research Trust and to Brig Klyce for their unstinting support of our research into panspermia.

J.T. Wickramasinghe
N.C. Wickramasinghe
W.M. Napier

Contents

Chapter 1

Overview

.....Microbiology may be said to have had its beginnings in the nineteen-forties. A new world of the most astonishing complexity began then to be revealed. In retrospect I find it remarkable that microbiologists did not at once recognise that the world into which they had penetrated had of necessity to be of cosmic order. I suspect that the cosmic quality of microbiology will seem as obvious to future generations as the Sun being the centre of the solar system seems obvious to the present generation.....
Fred Hoyle

1.1 Introduction

When Fred Hoyle made this prophetic statement at a public lecture in Cardiff on 15 April 1980, theories of panspermia were still considered heresy. Despite the inherent strength of the argument the hardened sceptic could not be easily swayed from the conventional wisdom of a purely terrestrial origin of life. Some three decades on the situation is rather different. Panspermia theories are now being discussed as a serious possibility for the origin of life on Earth, and indeed the very ideas that were hotly contested in 1980 are now sliding imperceptibly into the realms of orthodox science.

In recent years the limits of microbial life on the Earth have expanded to encompass an extraordinarily wide range of habitats: geothermal vents, the ocean floor, radioactive dumps and Antarctic soil, eight kilometres underneath the Earth's crust, to name but a few. The long-term survivability of bacteria has also been extended from 25–40 million years (Cano and Borucki, 1995) to a quarter of a billion years in the case of a bacterium entrapped in a salt crystal (Vreeland *et al.*, 2001). Such properties, particlularly in the case of extremophiles, are coming to be regarded as being of crucial relevance to astrobiology (Cowan and Grady, 2000).

The theory of panspermia (Hoyle and Wickramasinghe, 1981, 1982) does not address the question of a first origin of life, but only argues for its continuation once an origin is achieved. Starting from the premise that a *de novo* origin of life involves superastronomical improbabilities, two well-attested empirical facts are invoked to justify its case. Firstly, as stated earlier, microbes under appropriate conditions, have an almost indefinite persistence and viability. Secondly, given the right conditions and environments microbes can replicate exponentially.

Hoyle and Wickramasinghe sought over many years to identify interstellar and cometary dust with bacteria and their degradation products. Indeed the first identification of organic dust in space was made by Wickramasinghe (1974), and the first suggestion of cometary dust being organic, was made by Vanysek and Wickramasinghe (1975). An organic characterisation of the dust is now universally accepted although the original papers are rarely cited (Kwok, 2009). The complex organic character of cometary dust, including molecules that have a biological relevance, is used nowadays to argue that comets were important for bringing the organic primordial soup to Earth, although a purely terrestrial scheme for the origin of life is preferred thereafter. The alternative biological interpretation of cosmic dust was based on infrared spectroscopy, and required a third of the carbon in interstellar space to be tied up in the form particles that resembled bacteria and their degradation products (Hoyle and Wickramasinghe, 1982). Occam's razor cannot be used to discard this possibility so long as the efficiency of any competing inorganic process leading to similar spectra remains unresolved. Objections to panspermia invariably stem from a prevailing attitude of geocentricism. The idea of Earth-type microbial life pervading the Galaxy might seem absurd at first sight, but even more daring would be the proposition that such life, once it gets established anywhere, can be localised and confined.

Fred Hoyle (1915–2001). One of the most outstanding scientists of the 20th century, he made pioneering contributions to many different areas of astronomy — accretion theories, stellar evolution, cosmology and interstellar matter — to name but a few. His early work on stellar evolution led to his famous collaboration with Margaret and Geoffrey Burbidge and William Fowler on the synthesis of the chemical elements beyond helium in stars. Together with Bondi and Gold he developed the steady-state cosmological model. His collaborations with one of the present authors stretched over a full four decades. Even before his entry into the panspermia debate he was an early supporter of the modern view that extrasolar planets and life are ubiquitous — describing these ideas in fictional form for instance in the *Black Cloud* (Photo by Chandra Wickramasinghe).

1.2 Cometary Panspermia

Comets are known to have formed in the early stages of the condensation of the solar system. Cometary panspermia requires a small fraction of microorganisms present in the interstellar cloud from which the sun and planets formed to have retained viability, or to be capable of reactivation after being incorporated within newly formed comets. The fraction could be exceedingly small. The present-day Oort cloud contains some 100 billion individual comets and their total mass is comparable to the combined masses of the outer planets Uranus and Neptune. With one percent of the mass of the initial comet cloud being made up of interstellar dust the total number of 'graveyard bacteria' accommodated in a single comet would be some 10^{28}. A viable fraction as small as one part in 10^{18} would still yield some ten billion bacteria for each newly formed comet. If replication can occur within the comet, the previous history of destruction and inactivation in the interstellar medium becomes irrelevant, because of the capacity of even a single viable microbe to increase exponentially.

Water is a major component of comets, but it had been thought for a long time that this could only exist in the form of ice. However, the cloud from which the solar system formed contained free radicals and molecules from interstellar space as well as radioactive isotopes, including ^{26}Al. The processes of recombination of radicals and exothermic chemical reactions between molecules in newly formed comets would inevitably provide heat sources that would transiently melt their interiors (Hoyle and Wickramasinghe, 1978, 1979). Radioactive decays of elements such as ^{26}Al would be an additional — perhaps more important — energy source for preserving liquid interiors in primordial comets for timescales of the order of a million years (Wallis, 1980; Hoyle and Wickramasinghe, 1980, 1985). These isotopes were present because a nearby supernova dispersing such nuclides had exploded when the solar system formed.

As we shall show in Chapter 7 radioactive heat sources served to maintain a warm liquid interior in each one of 100 billion comets for the major part of a million years, and this was ample time for the minute surviving fraction of interstellar microbes to replicate and fill a substantial fraction of the volume of a comet. Cometary activity in the outer regions of the solar system 4 billion years ago, including collisions between cometary bodies, would have led to the expulsion of a fraction of regenerated bacteria back into interstellar space. A fraction of comets would also be deflected into the inner regions of the solar system, thus carrying microorganisms onto the Earth and other inner planets. The first life on the Earth 4 billion years ago would, according to this model, have been brought by comets.

1.3 History of Panspermia

The concept of panspermia, that life is ubiquitous within the Universe, has a long and chequered history. The Greek philosopher Anaxoragas stated in the 5th century BC that life-seeds were distributed everywhere in the cosmos, taking root wherever conditions became appropriate. Even much earlier in the second millennium BC Vedic philosophies of ancient India included the basic idea of panspermia. There are also ancient Egyptian inscriptions of a similar date that can be interpreted as depicting panspermia (Temple, 2007). However, the view that dominated Western philosophy for nearly two millennia is one generally attributed to the philosopher Aristotle. Aristotle came at the end of a long line of rational philosophers including Plato who enunciated the doctrine of spontaneous generation — life emerging from inorganic matter spontaneously under the right conditions. Thus Aristotle famously proposed that "fireflies emerge from a mixture of warm earth and morning dew", and down the ages there were many variants on same theme.

Spontaneous generation of life must have appeared entirely reasonable to observers of nature who had only their unaided eyes with which to observe the world. The invention of the microscope by Anton Leeuwenhock made possible a challenge to this doctrine at a microbial level. Even so, there were arguments that were raised about contamination whenever spontaneous generation was supposed to have been demonstrated. Thus Felix Pouchet in 1860 published a thesis in which he claimed was definitive proof of spontaneous generation. This was later shown to be the result of contact with "contaminated" air. It was Louis Pasteur in 1868 who carried out a series of carefully controlled experiments, for example on the souring of milk and the fermentation of wine, in which he showed convincingly that these processes are caused by ambient microbes, since when the experimental flasks were not exposed to air, no fermentation occurred. Pasteur claimed through his experiments to deal a "mortal blow" to the theory of spontaneous generation. It is ironic that he also appears to have laid the first experimental foundation for panspermia.

Pasteur's experiments showed that microorganisms are always derived from pre-existing microorganisms, an idea which was expanded upon by Lord Kelvin: *"Dead matter cannot become living without coming under the influence of matter previously alive. This seems to me as sure a teaching of science as the law of gravitation..."* The idea of interplanetary panspermia was discussed by Kelvin in 1871 and involved the transport of life-bearing rocks between objects within the solar system. It was his opinion that such transfers of life between the inner planets could occur sporadically as a consequence of asteroid and comet impacts.

William Thomson (Lord Kelvin) (1824–1907). Belfast born physicist based in Glasgow University. One of the most outstanding scientists of the 19th century who made many fundamental contributions to the mathematical analysis of electricity, to thermodynamics and to geoscience, making original quantitative attempts to find the ages of the Earth and Sun (Courtesy: Wikimedia Commons).

In 1874 the German physicist Hermann von Helmholtz made the following statement:

"It appears to me to be fully correct scientific procedure, if all our attempts fail to cause the production of organisms from non-living matter, to raise the question whether life has ever arisen, whether it is not just as old as matter itself, and whether seeds have not been carried from one planet to another and have developed everywhere where they have fallen on fertile soil...."

Hermann Ludwig Ferdinand von Helmholtz (1821–1894). German physician, physicist and philosopher who made fundamental contributions to electromagnetism, ophthalmic optics, acoustics, and mechanics. He popularised the idea that the universe is evolving towards a heat death (Courtesy: Wikimedia Commons).

In the beginning of the twentieth century this idea was expanded upon by Svante Arrhenius. His hypothesis (Arrhenius, 1903, 1908) that bacterial spores were transported across the Galaxy by the radiation pressure of starlight was met with much criticism. The long-term viability of the microbes subjected to the effects of UV and ionising radiation during transit was questioned. Although such criticisms continue to re-surface in modern times, upon closer examination they are found to be without foundation. We shall return to this point in later sections.

Claims of the presence of microbial fossils in the Martian meteorite ALH84001 in 1996 (McKay *et al.*) led to a revival of interest in Kelvin's version of panspermia — lithopanspermia as it has come to be known. However, the direct slingshot transfers of life-bearing rocks from the inner planets to the outer planets, satellites and beyond might represent a relatively inefficient form of panspermia. According to Melosh (1988) a solar system boulder has an extremely low capture probability into an Earth-like environment.

However, the finding of interstellar transport mechanisms (Napier, 2004; Wallis and Wickramasinghe, 2004) makes the likelihood of life-transfer through variants of a lithopanspermia process much greater.

Napier describes how life-bearing boulders are ejected from the Earth following a comet or asteroid impact. Eroded by zodiacal clouds into dust particles and transported due to the effects of radiation pressure, they are subsequently reincorporated into nascent protoplanetary systems as the solar system moves in its ~240 My period orbit around the centre of the galaxy, encountering molecular clouds and giant molecular clouds as it does so.

Svante Arrhenius (1859–1927). Swedish chemist. One of the founders of physical chemistry, who also developed ideas about the causes of ice ages and developed the idea of carbon dioxide as a greenhouse gas which predates much modern thinking on the subject. He proposed that life might drift between planets by the transport of spores (Courtesy: Wikimedia Commons).

1.4 The Ultraviolet Problem

Becquerel (1924) was among the first to argue on the basis of laboratory experiments that bacteria could not survive space conditions, particularly exposure to ultraviolet radiation. Deactivation and damage of naked space-travelling individual bacteria was thought to occur due to the ultraviolet light of stars, and such criticisms have persisted into modern times (Mileikowsky *et al.*, 2000; Horneck *et al.*, 1993). However it has been shown (Lewis, 1971) that under normal laboratory conditions microorganisms are not easily destroyed by ultraviolet. Instead they are frequently deactivated due to the dimerisation of pyrimidine bases, a reversible process in which no genetic information is lost. Dimerisation of bases distorts the DNA configuration and impedes transcription. Repair takes place by exposure to visible sunlight or by the operation of specialised enzymic systems.

Conditions prevailing in interplanetary or interstellar space — cryogenic in the absence of air and water — are found to be minimally damaging. Most relevantly, perhaps, we note that it is possible to shield microorganisms against ultraviolet light. Molecular clouds in the Galaxy can remove the glare of ultraviolet radiation and enable the growth of protective coatings around bacterial particles. Thin skins of carbonised material around individual bacteria, only 0.02 μm thick, would also provide a shield against damaging ultraviolet radiation (see Chapter 6).

1.5 Resilience of Bacteria

Damage from ionising radiation has been a more serious objection to panspermia, with claims that exposure to cosmic rays in space over hundreds of thousands of years would prove fatal to microorganisms. These criticisms are not borne out by direct experiments under appropriate conditions, and likewise overlook the huge replicative power of bacteria. According to the theory of cometary panspermia (Hoyle and Wickramasinghe, 1979, 1981), even the minutest rate of survival ($\sim 10^{-24}$) would ensure the propagation of microbial life across vast cosmic distances.

During an average residence time of 10 million years the cumulative radiation dose (at a very low flux) received by a bacterium in a typical location in interstellar space is estimated as $\sim 10^6$ rad (1 rad = 10^{-2} joules of energy absorbed per kg of matter). Ionising radiation dislodges electrons, causing bond breaks in the DNA and forms reactive free radicals. Although many terrestrial bacterial species may not survive this process, some almost certainly would. Under laboratory conditions, doses of 2 megarads (2 Mrad) delivered over minutes limited residual viability of *Streptococcus faecium* by a factor of a million (Christensen, 1964), but similar doses have had a negligible effect on cultures of *Deinococcus radiodurans* or *Micrococcus radiophilus* (Lewis, 1971).

It is difficult to estimate the dose of cosmic rays received by a naked bacterium in a typical location in interstellar space, over a fraction of a million years. Within the solar system it is possibly in the range 10–45 Mrad per million years and depends on the distance from the Sun and the phase of solar activity, highest at times near the peak of the solar sunspot cycle. Such doses are higher than the integrated doses that have been delivered within the laboratory, where the survival of bacteria is well-attested. Yet there is uncertainty as to whether the terrestrial experience of radiation susceptibility could be directly translated to interstellar conditions. Exposure to short pulses of high intensity radiation may be far more damaging than ultra-low intensities of radiation delivered over millions of years. Furthermore, in anaerobic conditions with low O_2 pressures the effects of ionising radiation are diminished. Reducing H_2O content has a similar effect since it is the oxidation of free radicals, in particular OH^- that causes over 90% of DNA damage. Cryogenic temperatures, which immobilize and prevent the diffusion of free radicals, lead to a similar outcome.

Another issue of particular relevance to lithopanspermia is the survival of microbes at high shock pressures. Rocks harbouring microbial life ejected from a planet following a comet impact would have experienced very high shock pressures. Mark Burchell and his collaborators have pioneered techniques for testing survivability under these conditions (Burchell *et al.*, 2004). Pellets made of ceramic material infused with microbes were loaded into gas guns and fired at speeds in excess of 5 km/s onto solid surfaces. For *Bacillis subtlis* the survival fraction was found to be ~10^{-5} at shock pressure of 78 GPa, with smaller, yet finite survival rates applying for higher impact pressures.

1.6 Extremophiles

A hundred years ago the survival of microbes under space conditions may have seemed improbable, indeed impossible. Recent developments in microbiology, however, have shown the existence of many types of extremophilic bacteria surviving in the harshest of terrestrial environments. The discovery of bacteria in conditions which may replicate those found in space has particular relevance to astrobiology.

Psychrophyllic microorganisms have been found to thrive in Antarctic permafrost (Karl *et al.*, 1999). Junge, Eicken and Deming (2004) have shown that some psychrophiles metabolise at temperatures below 100 K, perhaps even as low as 50 K. Thermophyllic bacteria were found to replicate in water heated to temperatures above 100°C in deep sea thermal vents (Stetter *et al.*, 1990). An astounding total mass of microbes also exists some 8 kilometres below the surface of the Earth, greater than the biomass at the surface itself (Gold, 1992). A species of phototrophic sulphur bacterium able to perform photosynthesis at low light levels, approaching total darkness (Overmann and van Gemerden, 2000) has been recovered from the Black Sea. Bacteria such as *D. radiodurans* have even been found to thrive within the cores of nuclear reactors (Secker, 1994), seemingly unaffected by radiation.

Thus microbiological research has revealed that some microorganisms are incredibly space-hardy. These properties are now regarded as being crucial to astrobiology (Cowan and Grady, 2000). As the exploration of our solar system continues we will surely find extraterrestrial bodies with environments analogous to many of the terrestrial environments where extremophiles are known to flourish (Cleaves and Chalmers, 2004).

1.7 The Discovery of Organics in Cosmic Dust

The amount of dust entering the Earth's atmosphere from interplanetary space is usually estimated to be about 20,000 tons per year (Love and Brownlee, 1993). The mean geocentric velocity of larger particles, visual or radar meteors, is about 40 km per second. If the small particles have this mean geocentric speed, an influx of 20,000 tons per year would imply an interplanetary density of about $\sim 10^{-22}$ gcm^{-3}.

The spectroscopic signature of a bacterium or bacteria-like material in the 2.9–3.5 μm spectral region was found in interstellar dust (Allen and D.T. Wickramasinghe, 1981), signifying the first clear demonstration that over a third of the carbon in the Galaxy was tied up in the form of organic dust grains (Hoyle *et al.*, 1982). The infrared spectral correspondence unquestionably put a strain on any alternative inorganic processes for producing 'bacteria-like' grains so efficiently on a galactic scale.

Another clear signal that may be indicative of biology is the λ = 2175A ultraviolet absorption feature of interstellar dust which accords well with biologically derived aromatic molecules (Hoyle and Wickramasinghe, 1977; Wickramasinghe *et al.*, 1989). Thus at least two striking spectroscopic features of dust seem to be consistent with living material being present everywhere in the Galaxy.

The existence of similar material in comets was also vindicated spectroscopically when the first infrared spectra of comet dust similar to interstellar dust were obtained at the last perihelion passage of comet Halley in 1986 (D.T. Wickramasinghe *et al.*, 1986). Other comets studied since 1986 led to further confirmation of these results.

With the launch of the Infrared Space Observatory (ISO) by ESA in 1995 a large number of unidentified infrared bands (UIBs) have been discovered in emission at well-defined wavelengths between 3.3 and 22 μm in a wide range of types of astronomical object. Whilst polyaromatic hydrocarbons (PAHs), widely presumed to form inorganically, are the favoured model for the UIBs, it is the case that no satisfactory agreement with the available astronomical data has been shown possible, particularly if we require the UIB emitters and the 2175A absorbers to be the same. Biologically generated aromatic molecules will be seen in Chapter 2 to provide a better explanation of this set of data. Investigations of Guillois *et al.* (1999) have pointed to a promising model based on coal (anthracite), which is a degradation product of biology.

More recently Caltaldo, Keheyan and Heyman (2004) have shown that aromatic distillates of petroleum, another biological product, exhibits correspondence with the astronomical diffuse infrared bands (UIBs) as well as the λ = 2175A ultraviolet absorption feature. As the resolving power of telescopes improves we will of course continue to discover more and more organics in molecular clouds.

In Chapter 2 we shall also interpret data from the lens galaxy SBS 0909+532 at $z = 0.83$ and also the far ultraviolet extinction of starlight in our Galaxy and in external galaxies, showing a biological provenance of both.

1.8 Comets

A comet taken in isolation is a rather insubstantial object. But the solar system possesses more than a hundred billion comets so that in total mass they equal the combined masses of the outer planets Uranus and Neptune, about 10^{29} g. If all the dwarf stars in the Galaxy are similarly endowed with comets, then the total mass of all the comets in the Galaxy, with its 10^{11} dwarf stars, turns out to be some 10^{40} g. This is very close to the mass of all the interstellar grains.

We know from observations of comet Halley near perihelion in 1986 that comets do eject organic particles, typically at a rate of a million tons per day. An independent analysis of dust impacting on mass spectrometers aboard the spacecraft *Giotto* also led to a complex organic composition that was generally consistent with the biological hypothesis (Wickramasinghe, 1993). Thus one could conclude from the astronomical data that cometary particles, just like the interstellar particles, are *spectroscopically* similar to biological material.

Recent collections of dust from comet Wild 2 in the NASA *Stardust* mission have revealed organic structures that could be interpreted as either prebiotic molecules or degradation products of biology (Sandford *et al.*, 2006). However, it should be stressed that the *Stardust* mission was planned in the late 1980s, before life in comets was considered an option. As a result, no life-science experiment was undertaken, and the particle collection strategy using blocks of aerogel did not permit the recovery of intact microbes following high-speed impacts (Coulson, 2009).

On July 4th 2005, NASA's *Deep Impact* probe collided with comet Tempel 1. The resulting ejecta were analysed using mass spectroscopy and an increased amount of organics was found to have been released after the collision (A'Hearn *et al.*, 2005). Evidence of clay minerals was also discovered, this material being unable to form without the presence of liquid water. Thus it appears that the pristine cometary interior of

Tempel 1 contains both organics and water. These matters will be discovered in more detail in Chapter 7.

1.9 The Origin of Life

Biological material contains about twenty different types of atoms, the most important being carbon, nitrogen, oxygen and phosphorus. The ultimate source of origin of these chemical elements is stellar nucleosynthesis — the process by which the primordial element H is converted first to He and thence to C, N, O and heavier elements in the deep interiors of stars (Burbidge *et al.*, 1957). So at the level of constituent chemical elements our origins can undeniably be traced to astronomy.

Charles Darwin who laid the foundations of evolutionary biology, never alluded to the origin of life in his 1859 book *On the Origin of Species* (1859). He had, however, thought about the problem and formulated his own position in a letter to Joseph Hooker in 1871 thus:

"....It is often said that all the conditions for the first production of a living organism are now present, which could ever have been present. But if (and oh! what a big if!) we could conceive in some warm little pond, with all sorts of ammonia and phosphoric salts, light, heat, electricity, &c., present, that a proteine compound was chemically formed ready to undergo still more complex changes; at the present day such matter would be instantly absorbed, which would not have been the case before living creatures were found."

Darwin's remarks provided the basic scientific framework for exploring the problem of abiogenesis throughout the 20th century and beyond. In the late 1920s A.I. Oparin and J.B.S. Haldane (Oparin, 1953) fleshed out Darwin's thoughts into the familiar 'Primordial Soup Theory', proposing that the atmosphere of the primitive Earth comprised of a reducing mixture of hydrogen, methane and ammonia and other compounds from which the monomers of life could be readily generated. Primitive 'lightning' and solar ultraviolet provided the energy to dissociate these molecules, and the radicals so formed recombined through a cascade of chemical reactions to yield biochemical monomers such as amino acids, nucleotide bases and sugars.

In the 1950s the classic laboratory experiments of Harold Urey and Stanley Miller (Urey, 1952) showed how organic molecules relevant to life might be formed by sparking mixtures of inorganic gases in a flask. The formation of the 'primordial soup' was then thought to be an easy step to comprehend. This led to the belief that life could be generated *de novo* as soon as the biochemical monomers were formed. The formation of the first fully-functioning, self-replicating living system with the potential for Darwinian evolution still remains an elusive concept, however.

It is generally conceded that the path from chemicals to self-replicating biology must progress through a sequence of organisational steps. The most popular contender for one such early stage is the RNA world. In this model nucleotides polymerise into random RNA molecules that lead to autonomously self-replicating macromolecules (ribozymes) without the need for an intermediate enzyme (Woese, 1967; Gilbert, 1986). Likewise, other contenders for prebiotic development include the 'Iron-sulphur world' (Wachtershauser, 1990), the 'PNA (peptide nucleic acid) world' (Bohler *et al.*, 1995) and the 'Clay World' (Cairns–Smith, 1966), the latter involving an inorganic clay system serving as the informational template. In view of the high abundance of silicon in the Galaxy the clay world model might well have a special role to play in a cosmic context, as we shall see in a later chapter. The transition from any of these intermediate systems to the final DNA–protein-based cellular life form is still in the realm of speculation.

1.10 Modern Advances

In recent years, modern developments which favour a cosmic connection for life have emerged. By measuring carbon isotope ratios in sedimentary rocks, Mojzsis *et al.* (1996) have detected evidence of microbial life on the Earth before 3.83 Gy. This corresponds to the end of the Hadean Epoch, a period of late heavy bombardment of the Earth by comets and meteorites when *in situ* development of a primordial soup would have been impossible to achieve. In addition Mojzsis *et al.* (2001) have discovered evidence of a possible Hadean ocean as early as 4.3 Gy, while Nisbet and Sleep (2001) have argued that such intense bombardment during this same period would have periodically led to the complete evaporation of oceans.

Therefore, whilst *in situ* prebiotic evolution may not easily occur under such hostile conditions, perhaps the cometary collisions themselves injected life onto the Earth during the period 4.0–3.83 Gy. One could speculate that the bacterial phyla that survived under such episodes of recurrent evaporation were thermophiles, found at the base of the phylogenetic tree. Such lifeforms would have had the ability to survive the harsh prevailing conditions, and from these lifeforms a very much larger set could evolve.

A requirement of panspermia models is that all the higher members of the tree of life came in the form of genetic components that could be assembled in response to the Earth's varying physical conditions. The perceived evolution in the phylogenetic tree (Woese and Fox, 1977) of terrestrial life within its branches, Archaea, Bacteria and Eukarya, then becomes a relic of the re-assembly process of cosmically-derived genes.

Recent studies of DNA sequence phylogenies show that sequences corresponding to higher life forms and multiple domains of life are present even at the very root of the 'tree of life' (Olsen and Woese, 1997). The search for convergence to a single 'Last Universal Common Ancestor' is thus proving elusive, and the search should more logically be directed to a 'Last Universal Common Community'. Tepfer and Leach (2006) have argued persuasively for an ensemble of genes preserved in the form of plant seeds (which may even be partially degraded) as a model for panspermia. This is consistent with the idea of a 'Last Universal Common Community' rooted firmly in the cosmos.

1.11 Protoplanetary Nebulae and Extra-solar Planetary Systems

Panspermia implies the transference of life between habitable locations widely separated in the cosmos. The requirement therefore is that habitable abodes of life must exist in abundance on a galactic scale. Do astronomical observations support the existence of planetary systems outside our solar system?

The starting point of all theories of solar system formation is a fragment of an interstellar cloud, a central condensation of which ends up as a star. Surrounding a newly-formed star in an embryonic planetary system is a flattened, rotating, protoplanetary disc within which planets eventually form. Whilst an outer shell of icy cometary bodies could be expected to surround a planetary disc, such a shell which is optically thin would be more difficult to observe using present day observational techniques. Some of the most active regions of star formation in our immediate vicinity are to be seen in the Orion molecular clouds. Hubble telescope images have revealed the presence of many protoplanetary discs which show nebulae edge-on. Two nebulae that have been actively investigated at both infrared and millimetre wavelengths are β-Pictoris, and TW Hydrae, the latter showing evidence of mm- or cm-sized accumulations of planetary material in the disc (Wilner, 2005). Injection of viable bacteria and spores into discs such as in Fig. 1.1 would lead to seeding with life from an external source (see Chapter 5).

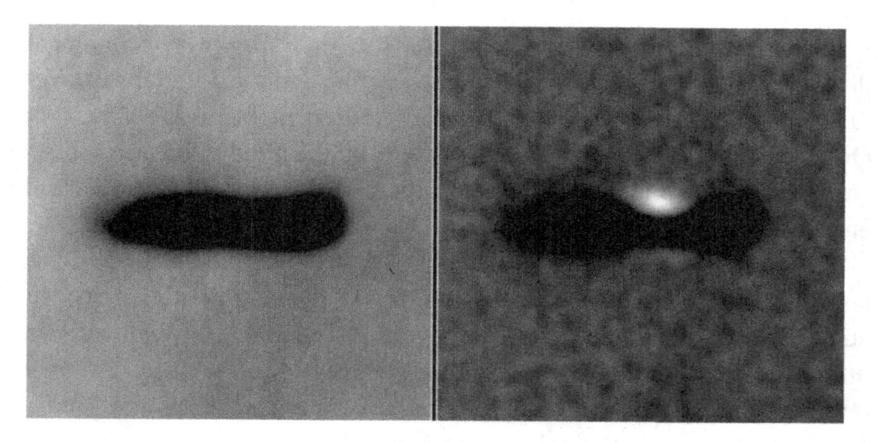

Fig. 1.1 Edge-on protoplanetary discs in the Orion Nebula (McCaughrean and O'Dell, 1996).

Theories of planet formation suggest that planets must be a commonplace occurrence, at least for stars of masses between 1 $M_\odot$ and ~10 $M_\odot$. Stability of planetary systems within a habitable zone around a parent star for timescales that are long enough for life to evolve is a separate issue which we shall not discuss here. The presence of any extrasolar planet on which microbial life can survive and replicate, would be of interest for panspermia and astrobiology.

Detection of extrasolar planets telescopically stretches modern astronomical technologies to the limit. If a planetary system like ours existed around our nearest neighbour α Centauri, its Jupiter counterpart would lie only 1" away from the central star, which would not be detectable against the brilliance of the central stellar disc even with our most powerful modern telescopes. Clearly indirect methods of extrasolar planet detection are required.

One such method that has been successfully deployed over the past decade is the use of the Doppler motions of central stars possessing Jupiter-mass planets. Viewed from the vantage of α Centauri our own Sun–Jupiter system would be detectable from dynamical effects. The star (Sun) and planet (Jupiter) would move around their common centre of mass with the period of the planet (11.9 yr), and this effect will be visible as a small regular Doppler wobble in the star's apparent path in the sky.

Marcy *et al.* (1995) pioneered the use of a Doppler technique to discover over 200 extrasolar systems, starting with the initial discovery of a planet around 51 Peg in the constellation of Pegasus (Mayor and Queloz, 1995; Marcy and Butler, 1996, 1998). Figure 1.2 sets out the characteristics of planetary systems discovered up until 2006, mainly employing the Doppler technique.

We note that the Doppler method has a selective bias to find Jupiter-sized planets in relatively close proximity to the central star. However, an exception to this rule has recently emerged with the deployment of an ultra sensitive planet-hunting spectrograph HARPS on the ESO 3.6 metre telescope. Using this instrument several Neptune-sized planets and an Earth-sized one have so far been discovered.

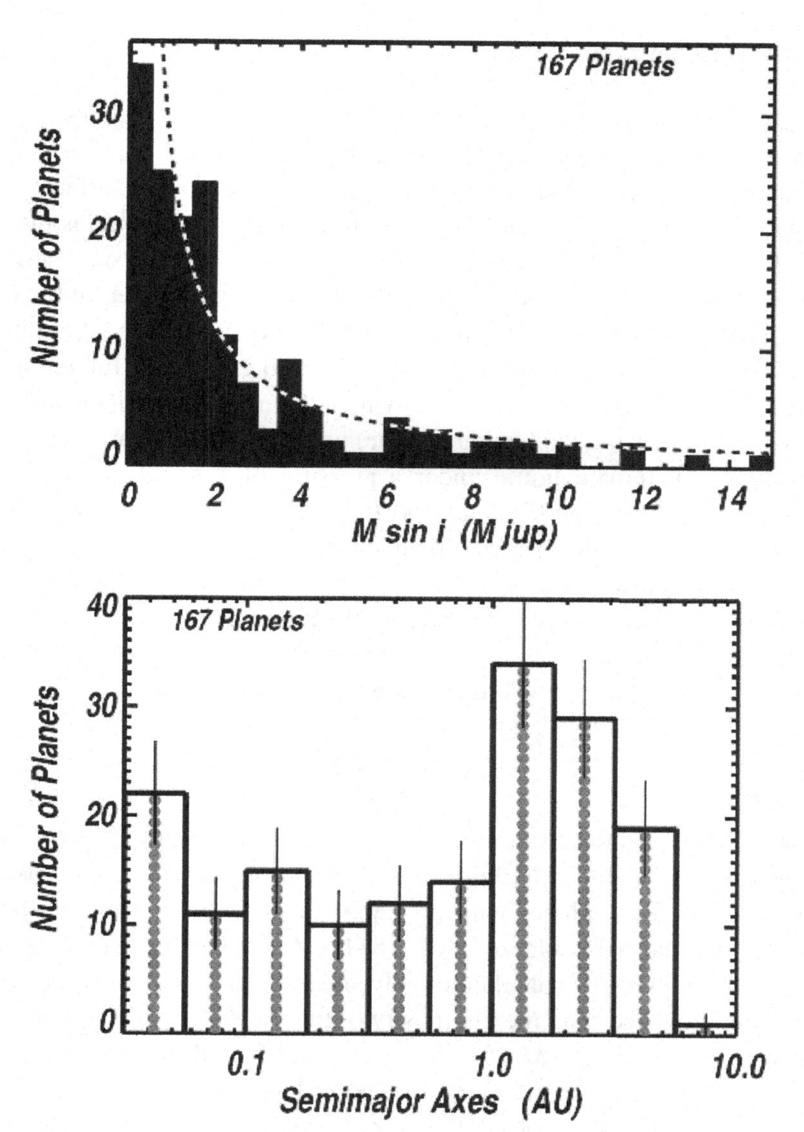

Fig. 1.2 Histograms showing the distributions of exoplanets discovered up to 2006; their minimum mass with $M \sin i < 15$ (upper panel), and their distance from star with $0.03 < a < 10$ in logarithmic distance bins (lower panel) (Butler *et al.*, 2006), i being the inclination.

1.12 Habitable Zone

A habitable zone around a star is defined as the range of radial distance in which a planet can maintain the conditions needed for life. This includes the requirement for liquid water at or near the surface, and ideally also a planet that can retain an atmosphere for timescales during which life can evolve. If the planet is too close to the star, surface temperatures would exceed the critical value for liquid water and if it is too far away the water will be in the form of solid ice. Another condition for a stable habitable planet is that it is not too close to a Jupiter-sized planet whose interactions could lead to it being perturbed inwards or outwards (away from the habitable zone) on timescales that are too short.

Water will remain liquid under a pressure of 1 bar (terrestrial sea-level pressure) between 0°C and 100°C. If complicating factors, such as the effect of an atmospheric greenhouse are ignored, a habitable zone for Earth-type life could be defined simply as the distance from a star where the effective temperature falls in the range 273–373 K. For a star of luminosity L the values r of the inner and outer bounds of the Habitable Zone (HZ) are given by the equation

$$L = 4\pi r^2 A\sigma T^4$$

where σ is the Stefan-Boltzmann constant, A is the Bond albedo of the planet and T is set equal to 373 K and 273 K respectively. In the case of the Sun and for a planet with A = 0.5, this yields a range of radial distance for the habitable zone of 0.8–1.5 AU. The relevant range for other stars can now be calculated easily since $r \propto L^{1/2}$. Furthermore since for Main Sequence stars the luminosity of the star, $L \propto M^{3.5}$, where M is the mass, we have $r \propto M^{1.75}$. A calculation along these lines is shown in Fig. 1.3. More realistic models of habitable zones around stars have been discussed by Jones, Underwood and Sleep (2005) and Franck et al. (2003) among others, taking account of factors such as an atmospheric greenhouse, and tidal locking.

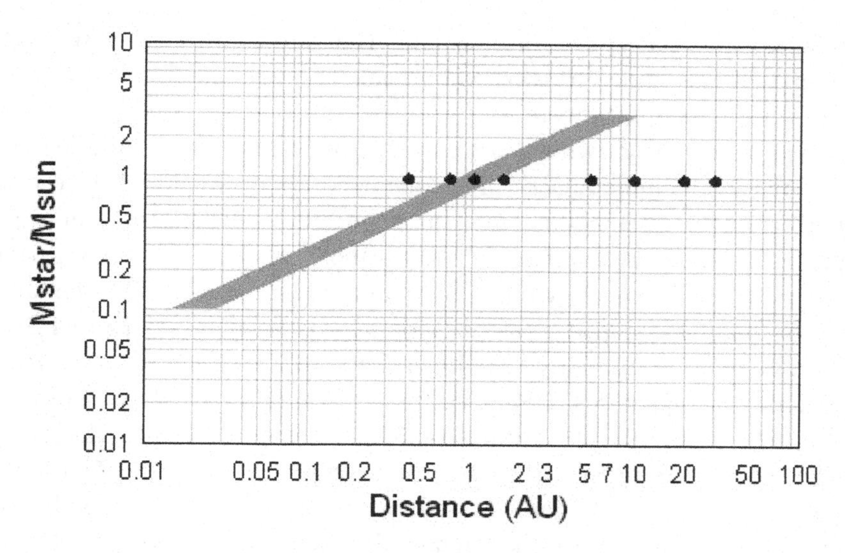

Fig. 1.3 Habitable zone (shaded area) calculated according to a simple model with bond albedo $A = 0.5$ for a prospective planet. The points represent the eight planets.

However, HZ models that are based only on the requirement supporting life on the surface of a rocky planet could turn out to be unnecessarily restrictive in our search for extraterrestrial life. Subsurface oceans, such as are almost certainly maintained in the Jovian satellite Europa through tidal heating, could define an additional habitable zone well outside the limits that are generally considered. Likewise Saturn's cloud-covered moon, Titan could also be warmed by a similar process, and the possibility of a habitable zone there cannot be ignored. In this book we shall also develop the idea that interiors of smaller bodies such as comets could also have interior domains heated by radioactive nuclides which would be congenial environments for microorganisms.

The COROT mission led by the French national space agency CNES was launched in December 2006 with one of its objectives being to investigate the existence of exoplanets using the transit method (The transit method detects minute dips in the stellar brightness, as the planet partially eclipses its disk). Around 120,000 Main Sequence stars will be studied by COROT during its 2½ year mission in the hope that many more exoplanets will be discovered. The Kepler Mission launched in 2008 will also study 100,000 Main Sequence stars continuously over 4 years using the transit method.

Confidence is growing that we may soon find that 'solar systems' like our own are commonplace. If so, the exceedingly improbable event of life's origins does not need to be replicated independently in every planetary abode that arises in the cosmos. According to panspermia theories the life legacy will be out there in space, ever-ready to be taken up by habitable comets, satellites or planets as they emerge. ESA's Darwin mission currently scheduled for launch in 2015 aims to detect biochemical signatures (H_2O, CH_4, O_3 etc.) on Earth-sized exoplanets by deploying a flotilla of orbiting 3-metre telescopes. New technologies have to be developed in time for this important project. Optimistic projections are that about 1% of stars in our neighbourhood have habitable planets. Since the average distance between stars is ~1.7 pc, the average distance between habitable planets would be ~8 pc.

Chapter 2

Cosmic Dust and Life

2.1 Introduction

Dust abounds in the Universe — in interstellar space, in intergalactic space, in external galaxies, in planetary systems, in comets and on planets like the Earth. The many conspicuous dark lanes and striations seen on a clear night against the backdrop of the Milky Way are gigantic clouds of sub-micron sized cosmic dust amounting in total mass to about a percent of the mass of the material that lies between the stars (see Photo 2.1). New stars, including planetary systems, condense from such clouds of interstellar dust, so its role in astrophysics is by no means trivial. Dust attenuates the ultraviolet light that causes the dissociation of molecules, thus permitting the formation of H_2 in dark clouds. The far-infrared radiation from dust removes the gravitational energy of collapsing clouds, assisting in the fragmentation of clouds and the formation of stars. Cosmic dust may even be connected with life itself.

The nature of interstellar dust has never ceased to spark fierce controversy amongst astronomers (Wickramasinghe, 1967; Hoyle and Wickramasinghe, 1991), a situation that has not significantly changed for nearly a century. In the 1930s the fashionable model of interstellar dust involved iron grains with typical sizes of a few hundredths of a micrometre, a model proposed by C. Schalen by analogy with the composition of iron micrometeoroids (Schalen, 1939).

In the mid-1940s the emphasis shifted to volatile grains of an icy composition, broadly similar to the particles that populate the cumulus clouds of the Earth's atmosphere. H.C. van de Hulst (1946) had argued that such grains would condense in interstellar clouds, and J.H. Oort and van de Hulst (1946) developed the theory of interstellar grain formation in considerable detail. By the close of the 1950s the dirty ice grain model was one of the standard paradigms of astronomy. Astronomers had ceased to worry about how these grains might be formed, but were only concerned with corrections that were required in order to take account of dust absorption and scattering for estimating distances, luminosities and temperatures of stars.

Photo 2.1 Interstellar clouds in the Horsehead Nebula in Orion (Courtesy: Anglo–Australian Observatory). Patches of obscuration show evidence of dense dust clouds blotting out background starlight; bright patches are clouds of hot ionised gas.

In 1960 Fred Hoyle and one of the present authors (NCW) re-opened the debate on the composition and formation of grains. It was argued that nucleation of ice particles in interstellar clouds where gas densities were typically in the range 10–1000 atoms per cubic centimeter was implausible, and attention was directed to astrophysical venues where much higher densities prevailed. Such venues included atmospheres of cool stars (Hoyle and Wickramasinghe, 1962), protoplanetary discs, supernova ejecta and comets – venues that were considered in quick succession as possible sites for the formation of grains (Hoyle and Wickramasinghe, 1962, 1968, 1970, 1979).

More recently, the idea of organic grains and biologically generated grain models was discussed (Hoyle and Wickramasinghe, 1991). In this chapter we discuss evidence supporting biological grain models, that is to say dust particles that have a biological provenance. It was indeed the discussion of organic/biologic grains in the 1970s that provided a major impetus for reviving theories of interstellar panspermia.

2.2 Constraints on Composition

There are clearly observational constraints on permissible models of interstellar dust. Firm upper limits to density follow from the available abundances of elements in interstellar clouds. If grains formed in cool star atmospheres, supernova ejecta or comets, their average elemental abundances in interstellar clouds must be consistent with the abundances of elements seen in the atmospheres of stars within these clouds. This must hold true in particular for the sun and the interstellar cloud from which it condensed. The currently accepted solar relative abundance data is set out in Table 2.1.

Table 2.1 The relevant relative abundances in the Sun (Allen, 2000).

Element	AW	Relative Abundance
H	1	2.6×10^{10}
O	16	2.36×10^{7}
C	12	1.35×10^{7}
N	14	2.44×10^{6}
Mg	24	1.05×10^{6}
Si	28	1.0×10^{6}
Fe	56	8.9×10^{5}
Al	27	8.51×10^{4}

Table 2.1 together with the estimate $\rho_H \approx 3 \times 10^{-24}$g cm^{-3} for the average hydrogen density in the solar vicinity, and the atomic/molecular weights of the various proposed materials leads immediately to Table 2.2.

Table 2.2 Upper limits of mass densities of various types of grain.

Species	Max Density (g cm^{-3})
H_2O (ice)	4.8×10^{-26}
C	1.9×10^{-26}
Fe	0.6×10^{-26}
$MgSiO_3$ (ice)	1.2×10^{-26}
SiO_2	2.6×10^{-26}
Organic/Biologic $(H_2CO)_n$	4.7×10^{-26}

The situation is, however, more complicated because the assumption that solar abundances exactly reflect grain composition could be questioned. For those elements that can form refractory dust, including carbon, magnesium, silicon and iron, a partial separation of gas and dust by radiation pressure effects might be achieved. A strict adherence to sub-solar abundances for overall grain composition is therefore less secure than might be imagined.

Further clues on the composition of dust may in principle be obtained from observations of interstellar absorption lines in the ultraviolet. This yields the gas phase densities of various chemical elements, and hence could reflect depletions relative to solar values, thus pointing to material locked up in grains. Whilst most metallic elements are found to be depleted in this way, the depletions increase with density, results for carbon and oxygen are still insecure. Apart from C tied up as CO the bulk of the remaining carbon is likely to be locked up in grains or, perhaps, interstellar comets.

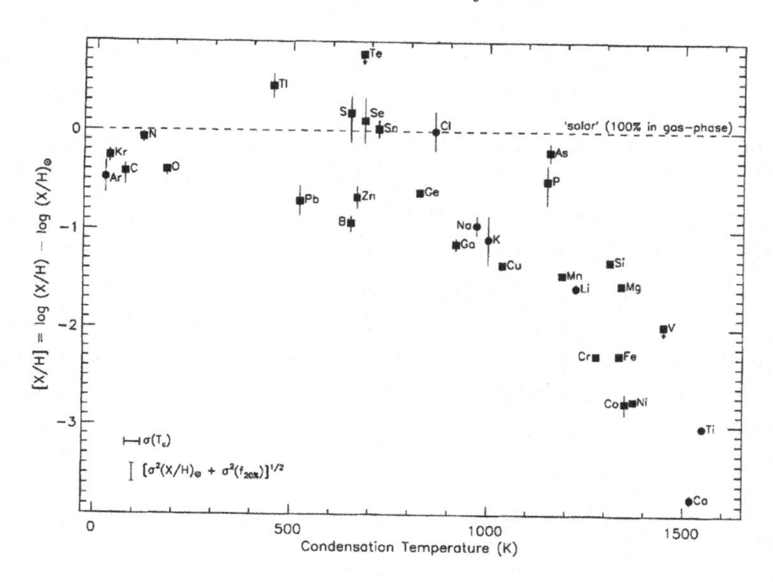

Fig. 2.1 Depletions of elements in diffuse interstellar clouds (Savage and Sembach, 1996).

2.3 Extinction by Spherical Particles

The problem of light scattering by a spherical particle is essentially one in classical electromagnetic theory and involves a solution of Maxwell's equations with appropriate boundary conditions on the sphere. Formal solutions were worked out independently by Mie (1908) and Debye (1909). The basic problem is as follows: a plane polarised electromagnetic wave is incident on a sphere of radius a, complex refractive index m. A long distance away the forward beam has lost a certain amount of energy. Part is *scattered* out of the forward beam and part *absorbed* by the sphere. The effective cross-sections for these processes are denoted $Q_{sca}\pi a^2$, $Q_{abs}\pi a^2$, where Q_{sca}, Q_{abs} are scattering and absorption efficiencies. The total cross-section for extinction is thus

$$Q_{ext}\,\pi a^2 = Q_{sca}\,\pi a^2 + Q_{abs}\,\pi a^2 \tag{2.1}$$

and we also define an albedo

$$\gamma = Q_{sca} / Q_{abs} \tag{2.2}$$

One of the objectives of the solution by Mie is to calculate Q_{sca}, Q_{abs} as functions of $x = 2\pi a / \lambda$, a dimensionless size parameter, and $m = n - ik$, where λ is the wavelength of the incident radiation, m being the complex refractive index which in general is dependent on λ.

Numerical codes are now readily available for computing Q_{sca}, Q_{ext}, Q_{abs} for spherical particles once the radius a, the wavelength λ and the optical constants n, k are given — the latter being in general functions of λ (Wickramasinghe, 1973). Examples of calculations of Q_{ext} for various values of m are given in Fig. 2.2.

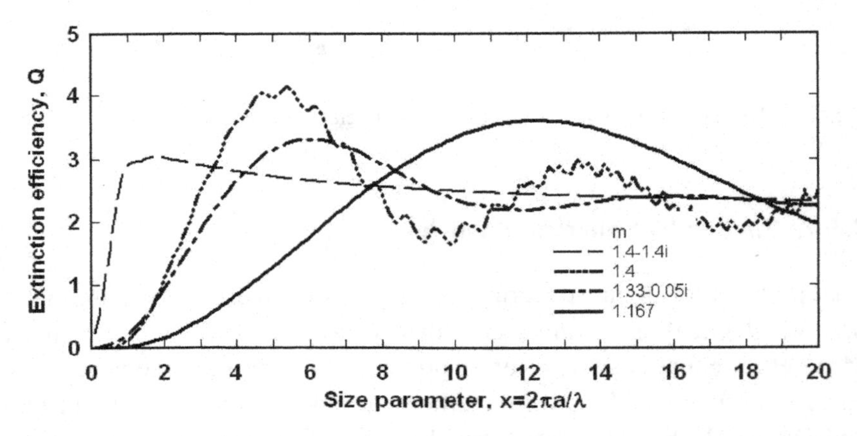

Fig. 2.2 Extinction efficiencies for spherical particles with various values of the refractive index.

It is customary to perform calculations for spherical isotropic grains in the belief that a random orientation of irregular-shaped particles can be considered equivalent (as far as extinction goes) to a spherical grain. In order to assess the plausibility of any theoretical dust model we require first to know the relationship between extinction cross-sections of individual grains and the astronomically determined extinction of starlight. This involves the solution of a simple equation of transfer which yields the change in magnitudes due to extinction

$$\Delta m(\lambda) = 1.086 N \pi a^2 Q_{ext} \tag{2.3}$$

where N is the column density of dust particles of radius a (Wickramasinghe, 1973). For a distribution of particle radii defined so that $n(a)da$ is the column density in the radius range $(a, a+da)$, Eq. (2.3) is modified to

$$\Delta m(\lambda) = 1.086 \int_0^\infty \pi a^2 Q_{ext}(a,\lambda)\, n(a)\, da \tag{2.4}$$

Since $\Delta m(\lambda)$ is proportional to $\langle Q_{ext}(\lambda) \rangle$ the normalised extinction determined observationally could be directly compared with a theoretical normalised extinction curve

$$\overline{Q(\lambda)} = \frac{\langle Q_{ext}(\lambda)\rangle - \langle Q_{ext}(\lambda_0)\rangle}{\langle Q_{ext}(\lambda_1)\rangle - \langle Q_{ext}(\lambda_0)\rangle} \tag{2.5}$$

where λ_0 and λ_1 are two reference wavelengths.

2.4 The Interstellar Extinction and Bacterial Dust

Early measurements of interstellar extinction at visual wavelengths by Stebbins *et al.* (1939) established a broad result that has survived unchanged. The visual extinction in magnitudes varies with wavelength approximately as $1/\lambda$ and the average value of the extinction coefficient of interstellar dust at 5000A in the solar vicinity amounts to about 1.8 magnitudes per kiloparsec. If ρ_g is the smeared-out density of dust and s is the specific gravity of grain material it follows that

$$\rho_g \cong 7.2 as \frac{10^{-22}}{Q_{ext}} \; g/cm^3 \tag{2.6}$$

where a is the average radius of a dust grain and Q_{ext} is the average the extinction efficiency at the wavelength 5000A.

Data on the wavelength dependence of interstellar extinction provides the most important direct discriminant of grain models. Until the mid-1960s observations over the near infrared to near ultraviolet waveband (0.9 μm to 3300A) could be made consistent with a wide range of models including ice grains with a typical radius of 0.3 μm and silicate grains of average radius 0.15 μm. Recent observations extending from the far infrared to the ultraviolet could be used to limit permissible grain models, however.

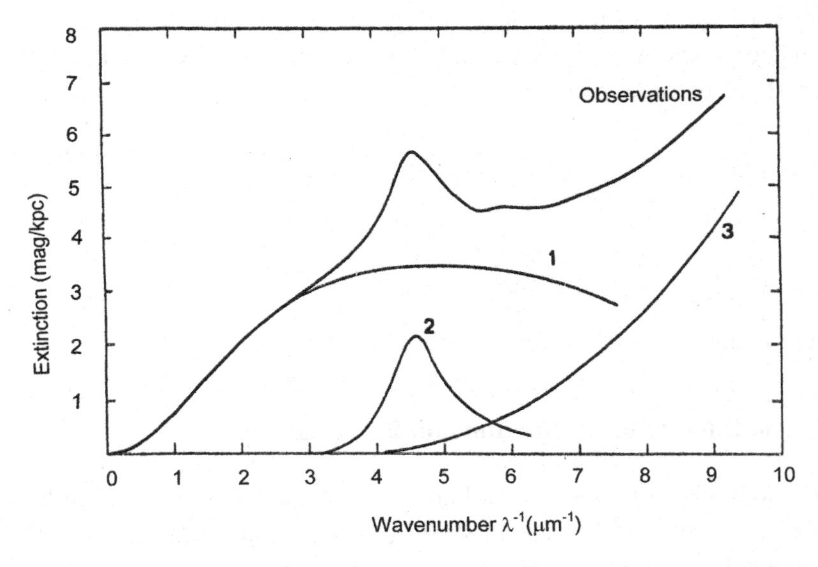

Fig. 2.3 Interstellar extinction curve in the Galaxy decomposed into 3 components (Hoyle and Wickramasinghe, 1991).

The smoothed-out mean extinction curve for the Galaxy from a compilation by Sapar and Kuusik (1978) is shown by the curve marked 'Obs' in Fig. 2.3. Modelling this data requires three basic components in general: (1) dielectric grains of average radii ~3000A; (2) graphite spheres of radius 0.02 μm and/or an unidentified molecular absorber to account for the peak of extinction at 2175A, and (3) dielectric grains of radii less than 200A. The 'dielectric' requirement for the components (1) and (3) permits any material with $n \sim 1.2$–1.6 and $k \sim 0$ in the visual and near ultra-violet spectral regions. Inorganic ices and silicates are possibilities, but infrared spectra reveal a different story as we shall see.

From the late 1960s onwards a consensus emerged among astronomers that the 2175A extinction peak was due to graphite particles (Wickramasinghe, 1967; Hoyle and Wickramasinghe, 1968). The requirement that the graphite had to be in the form of spherical isotropic grains with a single radius was tacitly accepted, although graphite is known to be highly anisotropic in its electrical properties, and graphite flakes rather than spheres would appear to be far more reasonable. Even modest departures from sphericity and deviations from a radius 0.02 μm for spheres of graphite moved the absorption peak away from the 2175A position. Calculations illustrating these points are shown in Figs. 2.4 and 2.5. Here we find a clear indication that a realistic graphite model for component (2) on the basis of isotropic graphite spheres of a precisely fixed radius (or mean radius) 0.02 μm is untenable.

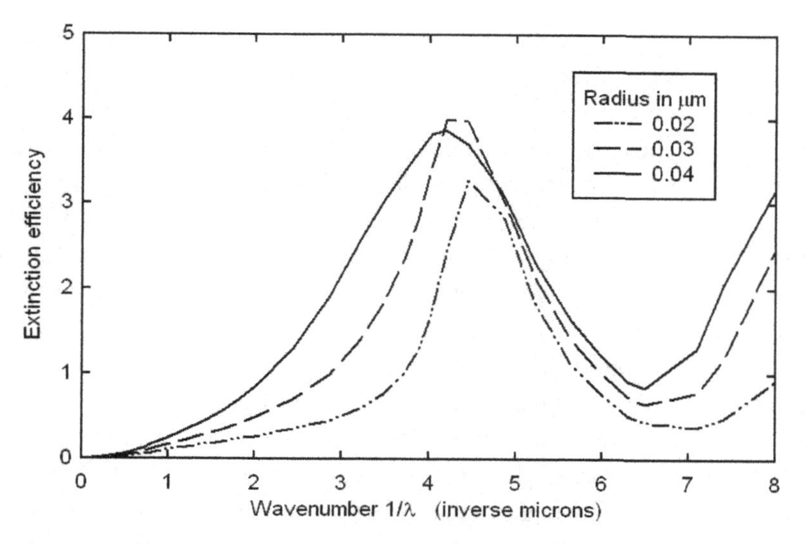

Fig. 2.4 Extinction efficiencies as function of wavenumber for graphite spheres of radii 0.02, 0.03 and 0.04 μm.

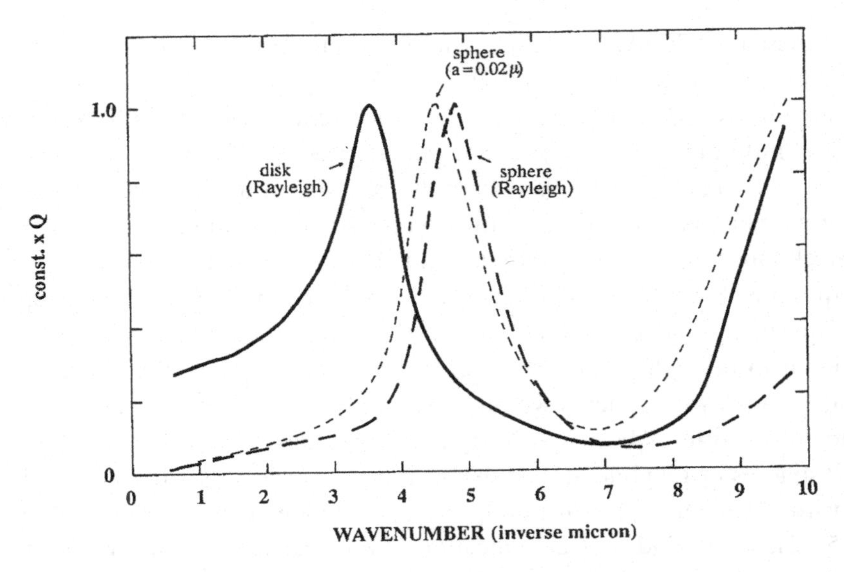

Fig. 2.5 Calculation of extinction curve for graphite in the form of Rayleigh scattering discs and spheres, compared with that for a sphere of radius 0.02 μm (Wickramasinghe *et al.*, 1992).

Notwithstanding these difficulties most contemporary research workers on interstellar extinction are firmly wedded to the idea of graphite spheres (Mathis, 1996; Draine, 2003; Zubko *et al.*, 2004).

The first molecular explanation of the 2175A absorption band (component 2) was proposed by Hoyle and Wickramasinghe (1977).

The fit of the astronomical data to a PAH model based on $C_8H_6N_2$ isomers is shown in Fig. 2.6, the model requiring a large fraction of interstellar carbon (not tied up in CO) to be in this form. PAHs are very much in vogue nowadays and are generally interpreted as interstellar condensates, although in our view they are much more likely to be generated via the degradation and destruction of life.

We now consider the interstellar extinction data in the restricted spectral region 1 μm to 3000A which is generally attributed to component (1) by most dust modellers. The points in Fig. 2.7 (lower panel) represent the mean galactic extinction law over this wavelength interval, the law applying to 'normal' interstellar dust, excluding HII regions and very dense molecular clouds. (In these exceptional regions either grain destruction or mantle growth could alter the mean size and compositional mix.) Although as we noted in Fig. 2.3 this extinction law obtained by K. Nandy (1964) has since been extended with respect to the wavelength base, the remarkable invariance of the relationship shown here (points in Fig. 2.7) continues to merit serious attention. On the basis of conventional models for component (1) of Fig. 2.3, the mean size of some type of dielectric grain (silicate or dirty ice) must be held constant in the diffuse interstellar medium to a high degree of precision. Whilst fixing grain sizes to within 20 or 30% throughout the Galaxy could be justified on physical grounds (e.g. response to radiation pressure), finer tuning of size would appear to be unreasonable.

Interstellar grains can have many different sources — cool stars, supernovae, bipolar flows from protoplanetary discs and comets. If cometary grains are biologically generated over several starforming cycles, a large fraction of the carbonaceous component of dust may be considered to have an origin in processes of bacterial replication.

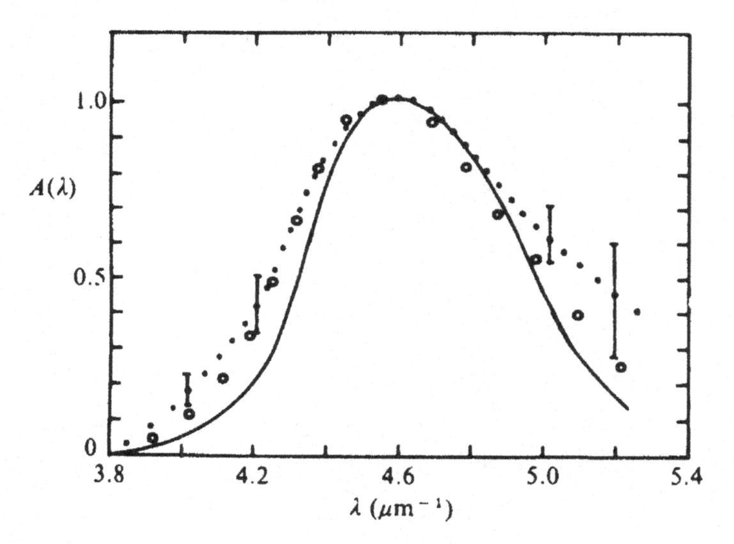

Fig. 2.6 Normalised absorptivity of $C_8H_6N_2$ isomers of a bicyclic aromatic hydrocarbon (solid curve) compared with astronomical extinction data as they existed in 1977 (points and dashed curve giving an average of the observations with notional error bars) (from Hoyle and Wickramasinghe, 1977).

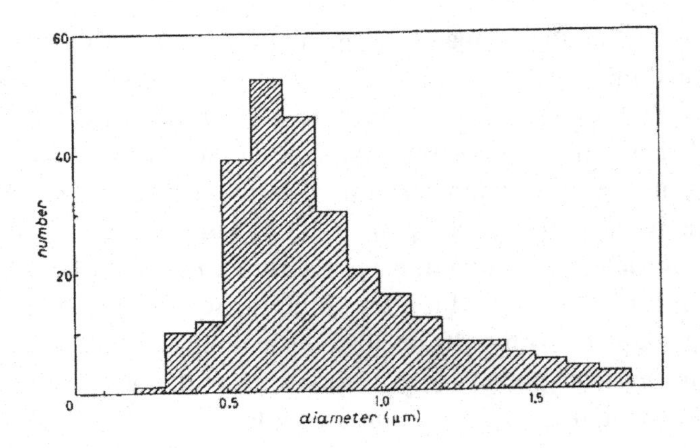

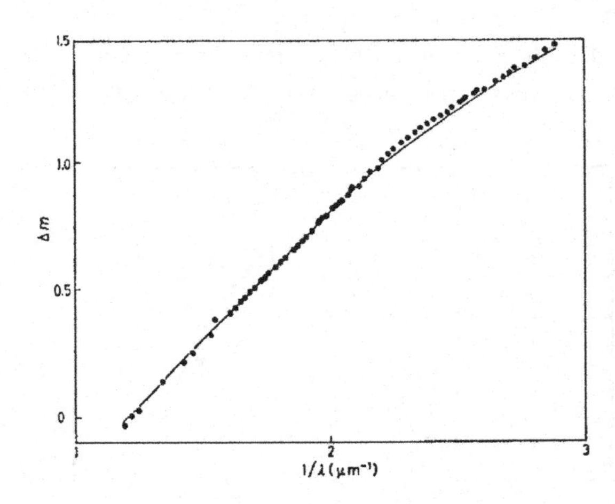

Fig. 2.7 *Lower panel*: Points represent the visual extinction data normalised to Δm = 0.409 at $1/\lambda$ = 1.62 μm^{-1} and Δm = 0.726 at $1/\lambda$ = 1.94 μm^{-1} (Nandy, 1964); curve is the calculated extinction curve for a size distribution of freeze-dried spore-forming bacteria with size distribution given by the histogram in the upper panel. The calculation uses the classical Mie theory and assumes hollow bacterial grains comprised of organic material with refractive index n = 1.4 and with 60% vacuum cavity caused by the removal of free water under space conditions.

Upper panel: Size distribution of terrestrial spore-forming bacteria as given in standard compilations of bacteriological data.

In a typical sample of such carbonaceous material the fraction of viable cells could be small, but their optical characteristics would not be expected to depart much from those of bacterial cells and spores as they leave an aqueous cometary environment and enter the near vacuum of interstellar space. With the evacuation of free water from a bacterium we estimate that 60% of its volume would be vacuum, and so the mean visual refractive index can be estimated as being n = 1.167 (Hoyle and Wickramasinghe, 1979).

Once the refractive index is fixed we have next to know the particle sizes in order to calculate extinction from Mie theory. Here too we have little leeway if we stick with biology as the source for most of the grains forming component (1). Figure 2.7 (upper panel) is a histogram of diameters of a representative sample of terrestrial spore-forming bacteria. The calculation of the extinction by such a distribution of particles all with refractive index n = 1.167 leads to the curve in the lower panel of Fig. 2.7 (*cf.* case n = 1.167 in Fig. 2.2). The amazing degree of correspondence between the observational points and the calculation gives confidence in the model.

Some bacteria contain hematite, with iron oxide molecules aligned in single domains. The susceptibility of these very strong little magnets is more than sufficient to produce orientation in the magnetic field of the Galaxy. So we have no trouble with explaining the observed few percent linear polarisation of starlight produced by a small subset of elongated bacterial grains.

Combining our model for component (1) with component (2) in the form of biological aromatics and component (3) in the form of dielectric nanoparticles of radii 50nm with various values of the refractive index n, we compute the normalised extinction curves shown in Fig. 2.8. The dashed and solid curves are respectively for refractive indices n = 1.5 and 1.4 of the nanoparticles (putative viruses or nanobacteria), contributing to 15% of the total extinction at 2175A. The points (crosses) are the astronomical observations of the total extinction of starlight (Sapar and Kuusik, 1978).

We note that the strikingly close fit to the astronomical data in the model of Fig. 2.7 (lower panel) is obtained with essentially no free parameters to fit. This feature must surely be considered a strong attraction of the biological model for interstellar extinction. The requirement in the biological model of desiccated bacteria with a composition and size distribution fixed only by biology itself, together with aromatic molecules resulting from degradation of biomaterial is to be contrasted with many models that demand a wide span of free parameters to choose from. Figure 2.9 shows one such model by Zubko *et al.* (2004) involving a mixture of silicate grains and graphite spheres (radius 0.02 μm) plus a component of PAH with an assigned, rather than measured, spectrum. Extensive reviews of attempts to explain all the data on interstellar dust in terms of conventional non-biological models are given by Draine (2003), Krishna Swamy (2005) and Kwok (2009).

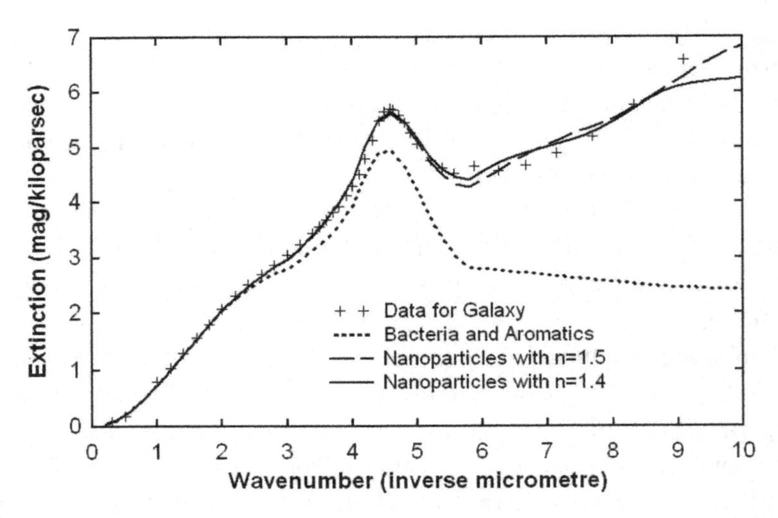

Fig. 2.8 The crosses represent the average interstellar extinction data (e.g. as compiled by Sapar and Kuusik, 1978, as well as later authors). The curves are for 3-component mixtures comprised of (a) hollow bacterial grains with average refractive index n = 1.167, (b) biological aromatics (115 individual spectra averaged by Wickramasinghe, Hoyle and Al-Jubory, 1989), and (c) dielectric particles of radii 50 nm with n = 1.5, 1.4. The dotted line represents the contribution from hollow bacteria plus their associated aromatics alone (References and credits in Hoyle and Wickramasinghe, 1991). The nanometric particles could represent silicates or nanobacteria.

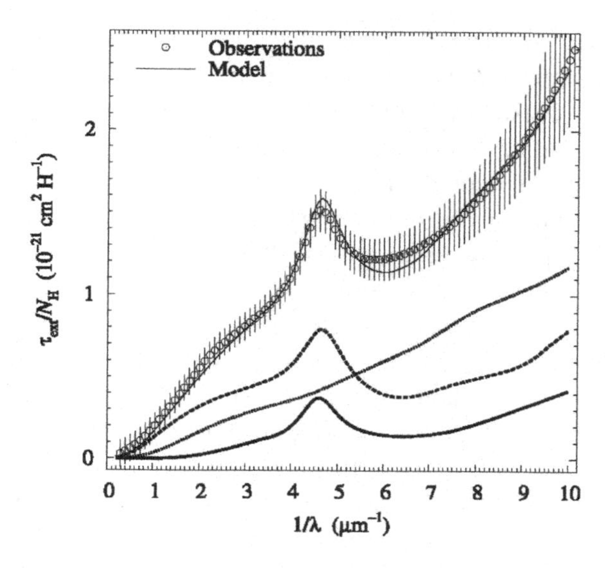

Fig. 2.9 A recent model including silicates, graphite particles of radii 0.02 μm and a hypothetical PAH — all with arbitrarily fixed sizes and weighting factors (Zubko *et al.*, 2004).

2.5 Infrared Evidence

Although the presence of organic solid material in dense molecular clouds was not in dispute by the end of 1978, evidence for their widespread occurrence in the general interstellar medium was still a matter for argument. The best chance of detecting organic dust in the ISM was to detect a clear 3.4 μm signal in sources of infrared radiation located near the galactic centre 10 kpc away, for which the visual extinction would be some 30 magnitudes. Observations with a poor spectral resolution that were available before 1978 already provided tentative evidence of the kind we sought, but decisive data became available only in 1981. In 1981 D.A. Allen and D.T. Wickramasinghe obtained a spectrum showing an unambiguous absorption feature in the Galactic centre infrared source GC-IRS7 with a characteristic shape indicative of an assemblage of complex organic material in the grains. The extinction over an 8 kpc path length at the centre of the 3.4 μm band was as high as 0.3 mag (Allen and Wickramasinghe, 1981).

It had been discovered earlier that the 3.3–3.5 μm absorption of desiccated biological cells was substantially invariant, both with respect to cell type (prokaryotes or eukaryotes) and to ambient physical conditions, including temperatures up to ~700 K (Hoyle *et al.*, 1982; Al-Mufti, 1984). The prediction then was that if grains were biological GC-IRS7 must not only possess absorption in this waveband, but it must also show absorption with an optical depth that was a rather precisely determined function of wavelength. This is exactly what was found when the first high resolution spectral data for GC-IRS7 was compared with the bacterial model. The modelling of the data was to all intents and purposes unique, and the fit shown in Fig. 2.10 implies that approximately 30% of all the available carbon in the interstellar medium is tied up in the form of material that is indistinguishable from bacteria.

Spectra over the 2.8–3.3 μm waveband very similar to the original spectrum of Allen and Wickramasinghe (1981) were published more recently by other observers (Okuda *et al.*, 1989, 1990; Pendleton *et al.*, 1994; Pendleton and Allamandola, 2002). The later attempts to measure the spectrum of GC-IRS7 have used more modern instruments. The generally favoured modern spectrum of GC-IRS7 is one attributed to Pendleton *et al.* (1994) which is reproduced as the points in Fig. 2.11. We see immediately that this spectrum differs from the original spectrum of Allen and Wickramasinghe (1981) (dashed line in Fig. 2.11) to the extent of an excess absorption over the 2.8–3.3 μm waveband that is generally consistent with the presence of water-ice. Our original conclusion concerning the *E-c*oli–GC-IRS7 opacity correspondence would remain valid provided we subtract an excess absorption near 3.4 μm, attributing it to spurious atmospheric water at the telescope site.

The far-infrared emission properties of dust in the Trapezium nebula and other sources also showed evidence consistent with a biological component. Although 10 and 20 μm emission bands had come to be generally accepted as implying silicate grains, deficiencies in the silicate models pointed to an organic component. Broad spectral features arising from C–O, C=C, C–N, C–O–C bonds are centred on wavelengths close to 10 and 20 μm and their contribution to the overall emission spectrum is inevitable. Figure 2.12 shows a comparison between the data for the

Trapezium nebula and a combination of biologically generated carbonaceous and siliceous material.

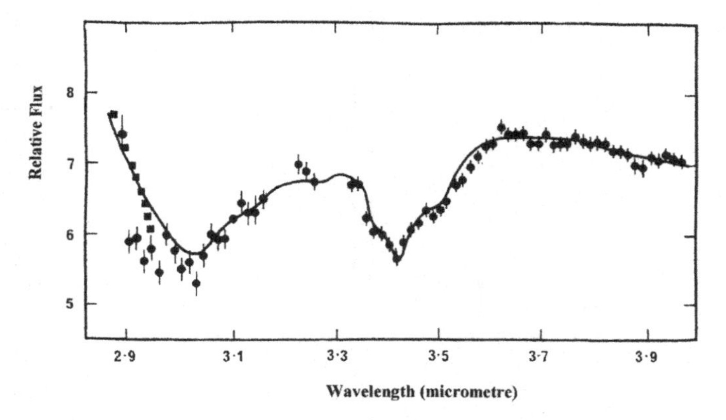

Fig. 2.10 The first detailed observations of the Galactic centre infrared source GC-IRS7 (Allen and Wickramasinghe, 1981) compared with earlier laboratory spectral data for dehydrated bacteria.

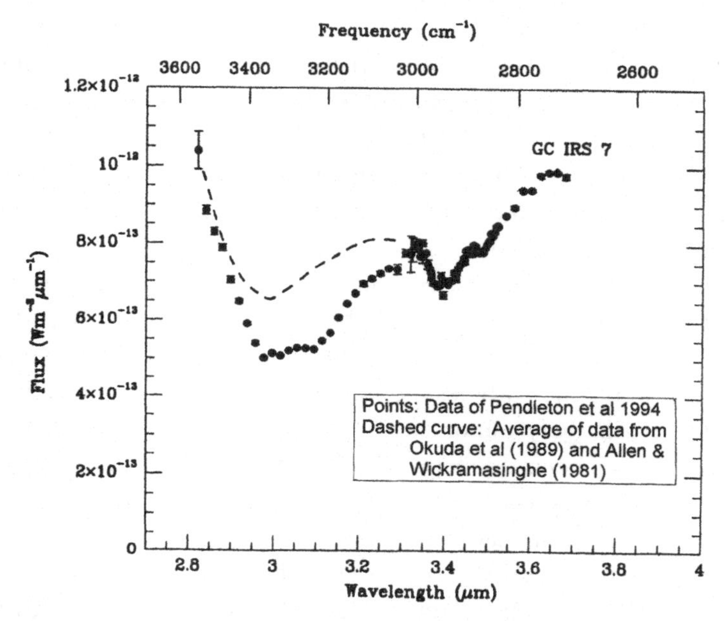

Fig. 2.11 High resolution data for GC-IRS7 (Pendleton *et al.*, 1994) (points). Dashed curve is the average relative flux values from the data of Allen and Wickramasinghe (1981) and Okuda *et al.* (1989).

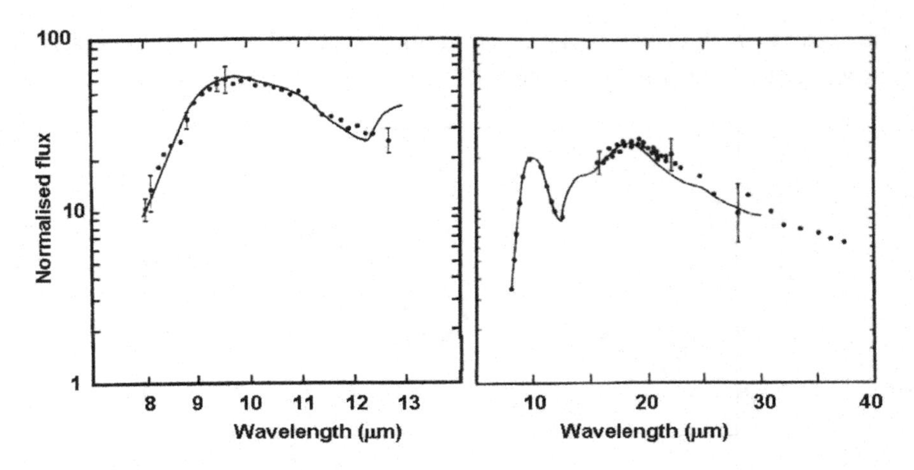

Fig. 2.12 Points are data for Trapezium nebula (Forrest *et al.*, 1975, 1976; Merril *et al.*, 1976). The curves show calculated emission behaviour of diatoms at 175 K.

2.6 Comet Dust and Biomaterial

There is growing evidence to support the view that the composition of dust in comets is very similar to that of interstellar dust, a claim originally made by Vanysek and Wickramasinghe (1975). Since the exploration of comet Halley in 1986 the infrared spectra of cometary dust have also been found to be consistent with a biological grain model, as shown for instance in Fig. 2.13.

As we pointed out earlier, such fits of biological systems to astronomical spectra do not imply the absence of silicates and other inorganic materials as constituents of grains. Indeed it would be surprising if inorganic particles did not exist, although the indications are that they are relatively minor components on the whole. One would perhaps expect planetary systems to be the places where mineral silicates abound, and this is indeed the case for the solar system where the dust in the inner regions is dominated by silicates.

As with the introduction of every new observing technique the use of ISO (Infrared Space Observatory), launched by ESA on 17 November 1995, provided new opportunities for testing astronomical theories. Spectral features near 19, 24, 28, and 34 μm observed by ISO have been attributed to hydrated silicates, such as in several protoplanetary discs, and in comet Hale-Bopp (Crovisier *et al.*, 1997; Waelkens and Waters, 1997). The uniqueness of some of these assignments is still in doubt, but even on the basis of a silicate identification of the principal infrared bands, such material could make up only some few percent of the mass of the dust, the rest being Trapezium-type grains, which also, according to our interpretations, (see Fig. 2.12) may be largely organic. This appears to be the case for the infrared flux curve of comet Hale-Bopp, obtained by Crovisier *et al.* (1997) when the comet was at a heliocentric distance of 2.9 AU. The jagged data curve in Fig. 2.14 may at first sight imply an overwhelming dominance of olivine grains. But our detailed modelling showed otherwise.

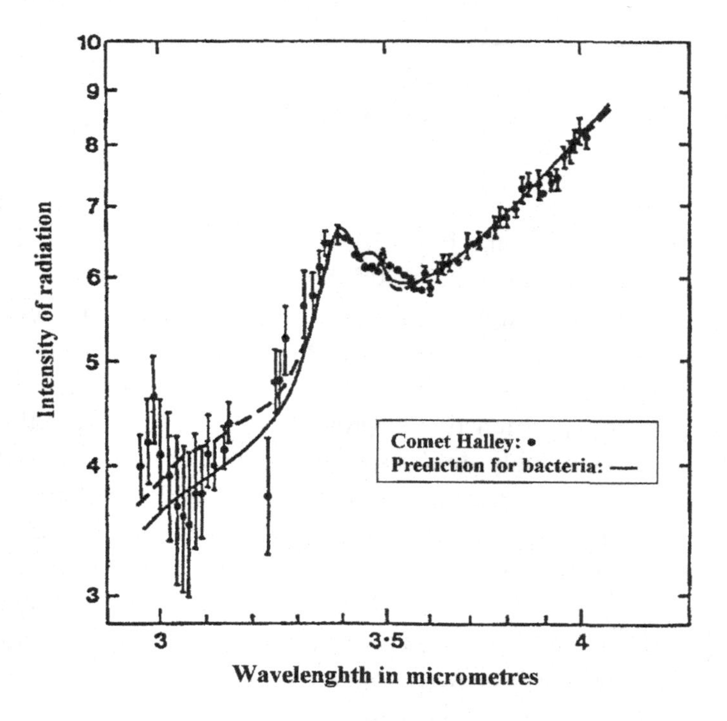

Fig. 2.13 Emission by dust coma of comet Halley observed by D.T. Wickramasinghe and D.A. Allen on March 31, 1986 (points) compared with bacterial models. Calculated curves are from Hoyle and Wickramasinghe (1991).

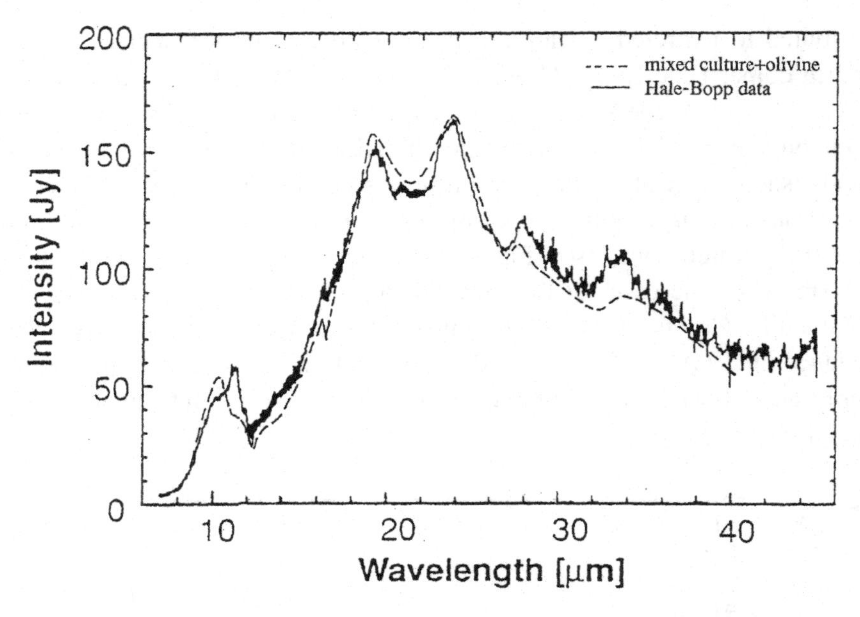

Fig. 2.14 Spectral feature near 3.4 μm in several comets after 1986.

The dashed curve in Fig. 2.14 is for a mixed culture of microorganisms containing about 20% by mass in the form of diatoms. Olivine dust, which has a much higher mass absorption coefficient than biomaterial, makes up only 10% of the total mass in this model (Wickramasinghe and Hoyle, 1999). It is interesting to note that the far-infrared emission spectrum of the protoplanetary disc around the star HD100546 has a strikingly similar spectrum to that of comet Hale-Bopp. A comparison of the two flux curves is shown in Fig. 2.15 (Malfait *et al.*, 1998). The difference in slopes of the two spectra between 20 and 40 μm is due to the dominance of lower temperature grains, but as in the case of Hale-Bopp the ratio of masses of crystalline silicates to biological material is likely to be low. The characteristic infrared features of crystalline silicates always tend to overwhelm the emission on account of their much higher intrinsic emissivities.

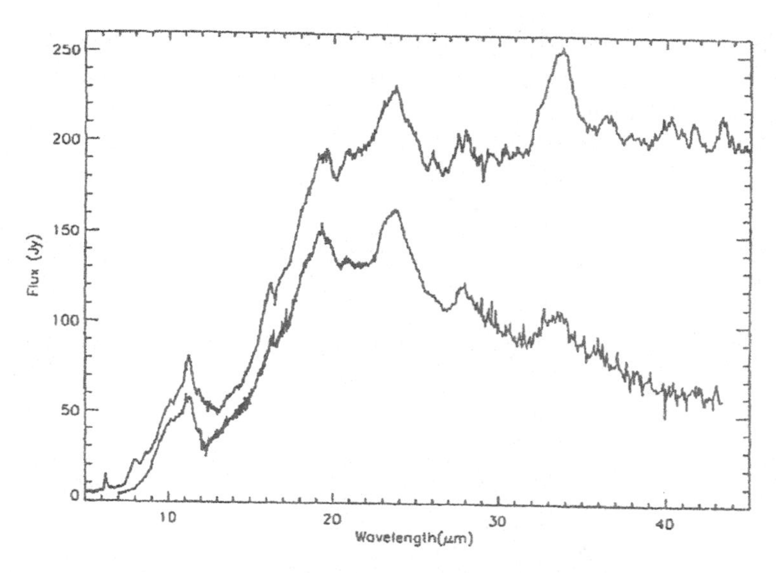

Fig. 2.15 Comparison between the spectrum of HD100546 (top curve) with the corresponding spectrum from comet Hale-Bopp (bottom curve) (Malfait *et al.*, 1998).

2.7 The Identification of PAH and Biological Aromatics

The distribution of unidentified infrared bands (UIBs) between 3.3 and 22 μm is almost identical in their wavelengths in very different emission sources, more or less irrespective of the ambient conditions. Most recent studies of UIBs for a large number of galactic and extragalactic sources have been obtained using the Spitzer Space Telescope (Smith *et al.*, 2007). A typical selection of such spectra is shown in Fig. 2.16.

Whilst PAHs (Polyaromatic Hydrocarbons), presumed to form inorganically, are the favoured model for the UIBs, no really satisfactory agreement with available astronomical data has thus far been shown possible for abiotic PAHs (Hoyle and Wickramasinghe, 1991). This is a particularly serious problem if we require the UIB emitters and the 2175A absorbers to be the same. The latter requirement is of course necessary because it is the staright energy absorbed in the ultraviolet band that is being re-emitted as UIBs in the infrared.

The Infrared Space Observatory (ISO) has also produced an impressive set of data on infrared emission bands, in both galactic and extragalactic sources.

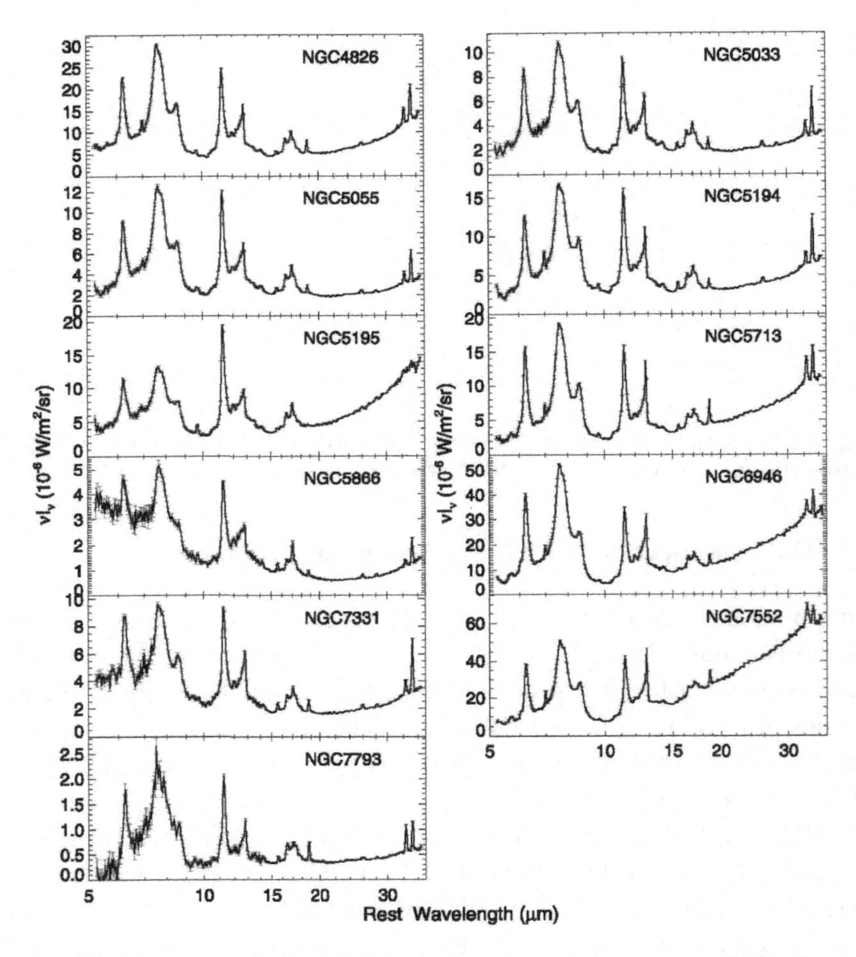

Fig. 2.16 Recent compilation of Spitzer telescope emission spectra of a variety of galactic sources showing PAH emissions (Smith *et al.*, 2007).

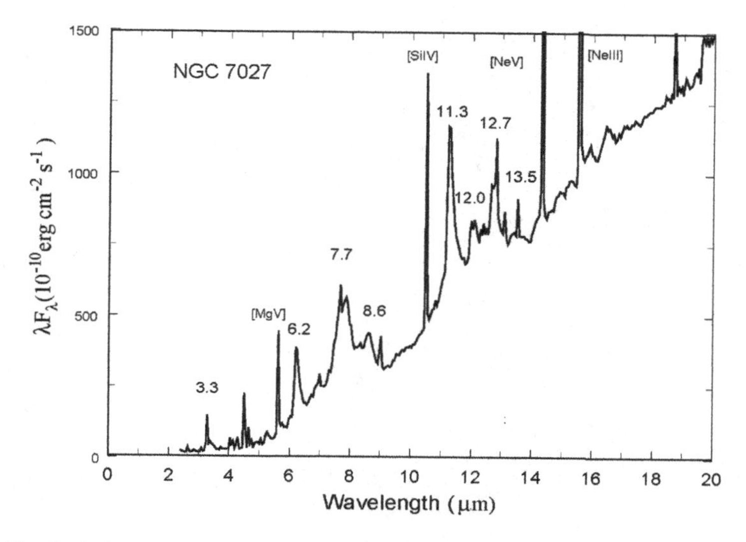

Fig. 2.17 Emission spectrum of planetary nebula NGC 7027 obtained by ESA ISO satellite.

Figure 2.18 shows a particularly interesting case where the spectrum of the Antennae galaxies at some 63 million light years distance has been observed to have infrared emissions that match the laboratory spectrum of anthracite (Guillois *et al.*, 1999). Anthracite being a product of biological (bacterial) degradation is again indicative of biological particles, in this case at great distances from the Milky Way.

More recently Caltaldo, Keheyan and Heyman (2002) have shown that aromatic distillates of petroleum, another biological product, exhibit correspondences with the astronomical diffuse infrared bands (UIBs) as well as the λ2175A ultraviolet extinction feature. Figures 2.19 and 2.20 show the results of Cataldo *et al.* (2002) for petroleum distillates. A tabulation of the principal infrared bands for comparison with the data in Figs. 2.18 and 2.19 is set out in Table 2.3 (e.g. Gezari *et al.*, 1993). The column for the biological aromatic ensemble is from the synthesised laboratory spectra obtained by Wickramasinghe *et al.* (1989, 1990).

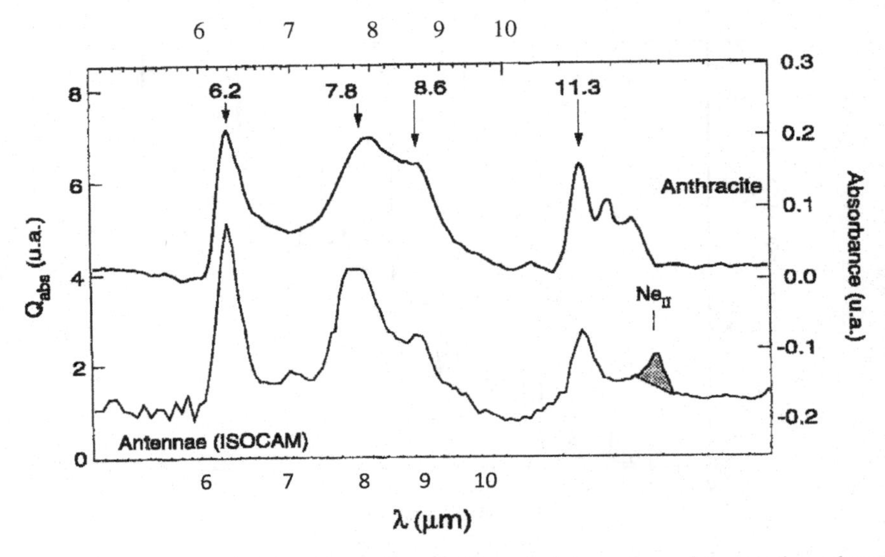

Fig. 2.18 Spectrum of anthracite compared with emission spectrum for an entire galaxy (Guillois *et al.*, 1999).

Table 2.3 Distribution of principal infrared emission wavelengths.

Principal UIBs:	Biological aromatic ensemble: 115 aromatics	Aromatic distillate extracted from petroleum	Anthracite
3.3	3.3		3.3
3.4	3.4	3.4	3.4
	3.6	3.5	
	5.28		5.2
6.21	6.21	6.2	6.2
			6.7
6.9	6.9	6.8	
7.2	7.2	7.2	7.2
7.7	7.7	7.6	7.8
8.6	8.6	8.6	8.6
	8.9	9.7	
		10.4	
11.3	11.21	11.5	11.3
12.2	12.14	12.3	
13.3	13.3	13.8	13.5

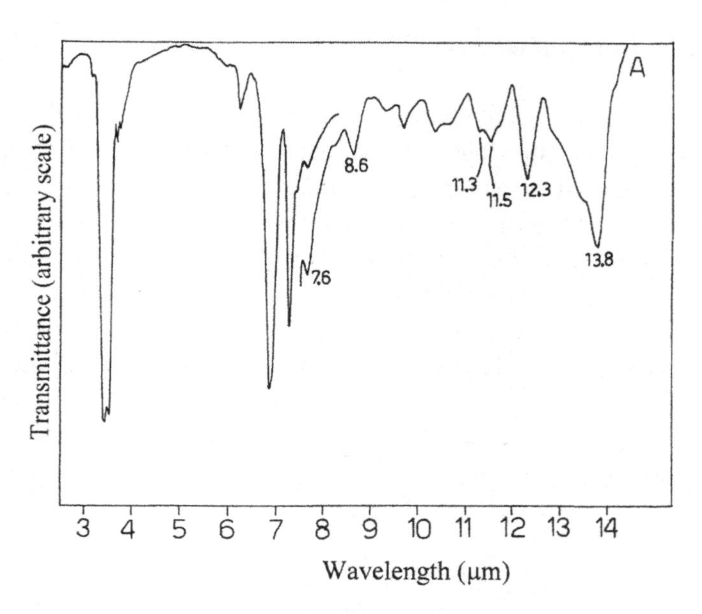

Fig. 2.19 Infrared spectrum of aromatic distillate from petroleum (Cataldo *et al.*, 2002).

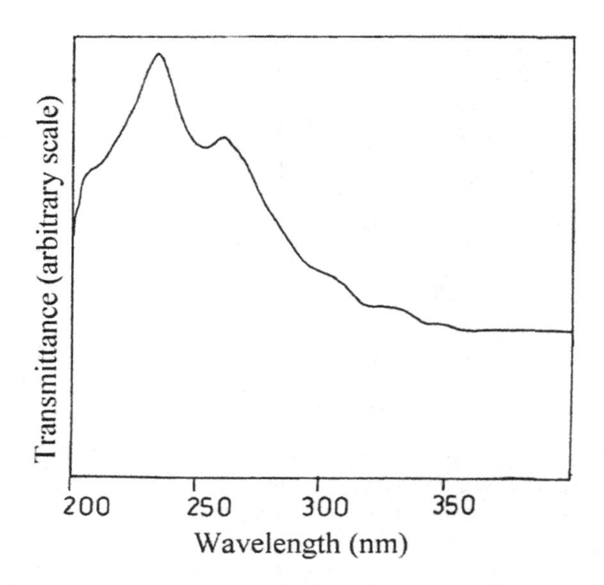

Fig. 2.20 Ultraviolet spectrum of aromatic distillate from petroleum (Cataldo *et al.*, 2002).

2.8 Other Spectral Features

The most dramatic recent discovery relating to astronomical aromatic molecules is a conspicuous 2175A band in the lens galaxy of the gravitational lens SBS0909+532 which has a redshift of $z = 0.83$ (Motta *et al.*, 2002). The extinction curve for this galaxy is reproduced in Fig. 2.21, with the dashed curve representing a scattering background attributed to hollow bacterial grains.

The excess absorption over and above a pure scattering background (dashed curve) is normalised to unity at the peak and plotted as the points with error bars in Fig. 2.22. The curve in this figure shows the absorption of biological aromatic molecules similarly normalised (Wickramasinghe *et al.*, 1989).

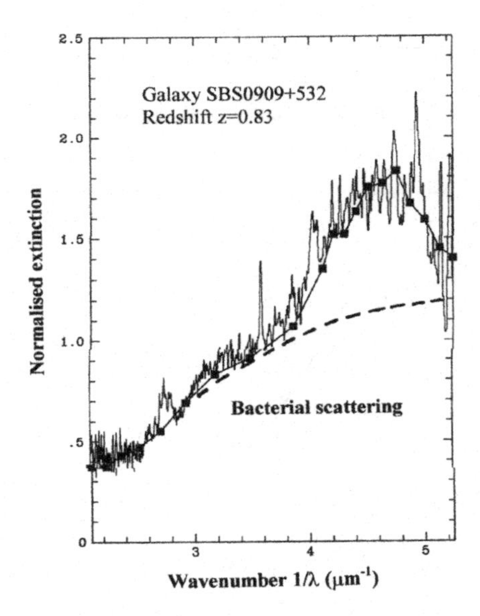

Fig. 2.21 The continuous line is the extinction curve for the gravitational lens galaxy SBS0909+532 excluding well-defined spectral lines due to MgII, CIII and CIV (Motta *et al.*, 2002). The dashed curve is the scattering background attributed to bacterial particles.

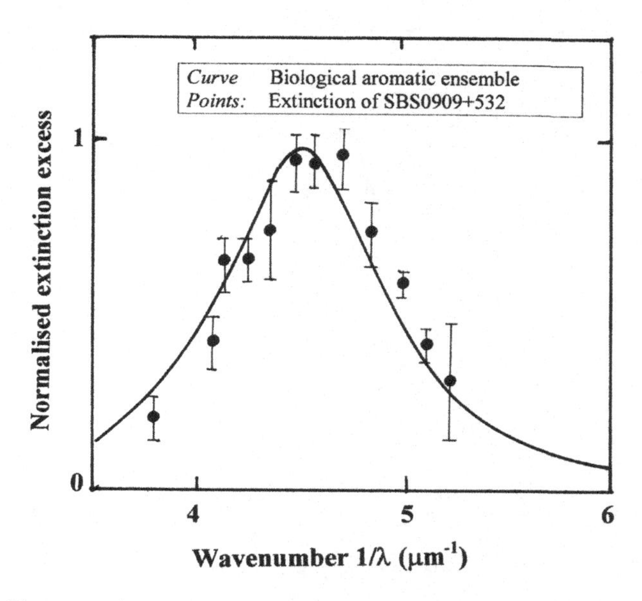

Fig. 2.22 The curve is the normalised absorption coefficient of an ensemble of 115 biological aromatic molecules. The points are observations for the galaxy SBS0909+532 due to Motta *et al.* (2002), representing the total extinction shown in Fig. 2.21, from which an underlying bacterial scattering component has been subtracted. The match between the data points and the theoretical curve shows that biological aromatics are a viable explanation of an absorption feature in a galaxy located at a redshift of $z = 0.8$. This was an epoch when the universe was nearly half its present radius.

A more detailed model with calculations involving a mixture of hollow bacterial grains, aromatic molecules and nanobacteria is shown in Fig. 2.23.

The correspondence between the astronomical data and the models in Fig. 2.23 can be interpreted as strong evidence for biology or biotic material at redshifts $z \approx 0.83$, that is up to a distance $D \approx cz/H \sim 2.5$ Gpc, assuming a Hubble constant of 100 km/s per Mpc. The new observations are consistent with the spread of microbial life encompassing a significant fraction of the radius of the observable universe. We shall discuss mechanisms by which microbial life could spread over cosmological distances in Chapter 9.

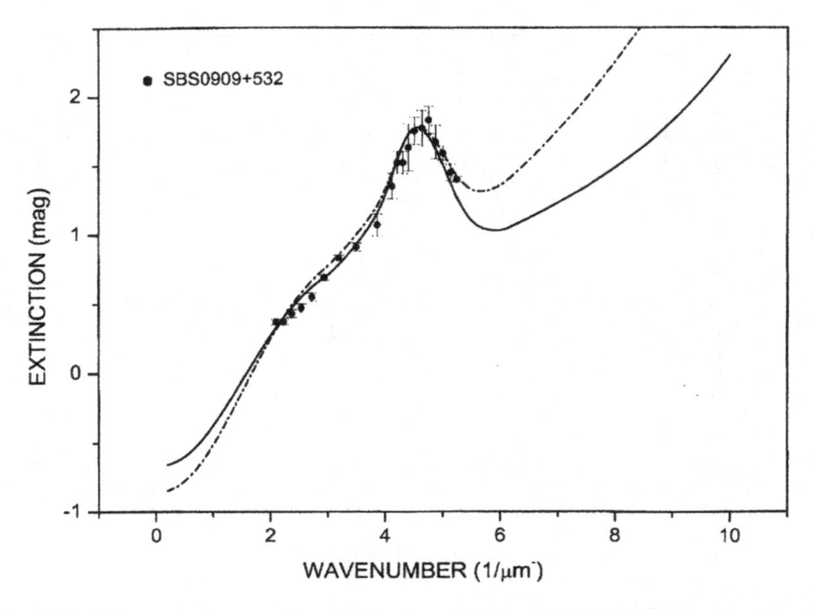

Fig. 2.23 Points are the normalised extinction data for SBS0909+532, the two curves are for a mixture of hollow bacteria, aromatic clusters and nanobacteria. The dashed curve has twice as much nanobacterial mass than for the solid curve.

2.9 Fluorescence

Biologically derived ensembles of aromatics have also been shown to match other astronomical datasets. The so-called extended red emission of interstellar dust, showing as a broad fluorescence emission band over the range 5000–7500 A is matched with biological pigments (Hoyle and Wickramasinghe, 1996; Wickramasinghe *et al.*, 2002a, b). The diffuse interstellar absorption bands in optical stellar spectra, particularly the 4430A feature, also have possible explanations on the basis of molecules such as porphyrins (Hoyle and Wickramasinghe, 1979; Johnson, 1971).

Extended red emission has been observed in planetary nebulae (Furton and Witt, 1992), HII regions (Sivan and Perrin, 1993), dark nebulae (Matilla, 1979) and high latitude cirrus clouds (Szomouru and Guhathakurta, 1998) in the Galaxy as well as in extragalactic systems (Perrin, Darbon and Sivan, 1995; Darbon, Perrin and Sivan, 1998). This phenomenon has a self-consistent explanation on the basis of fluorescence of biological chromophores (pigments), e.g. chloroplasts and phytochrome. Competing models based on emission by compact PAH systems are not as satisfactory, as is evident in Fig. 2.24. Hexa-peri-benzocoronene is one of a class of compact polyaromatic hydrocarbons that have been discussed in the astronomical literature. However, the width and central wavelength of its fluorescent emission leave much to be desired.

We have already mentioned that cometary dust and interstellar dust have generic and compositional similarities. Besides extended red emission from galactic and extragalactic sources of dust there is also evidence of 'excess redness' in objects within the solar system (Tegler and Romanishin, 2000). Some of the Edgeworth–Kuiper belt comets were found to have some of the reddest colours of any objects of the solar system. For the objects listed by Tegler and Romanishin we have taken B–V and V–R colours and calculated a reflectivity ratio given by

$$\frac{R(\lambda_R)}{R(\lambda_B)} = 10^{0.4\Delta(B-R)} \tag{2.7}$$

with $\lambda_R = 6420\text{A}$, $\lambda_B = 4400\text{A}$.

A redness factor $f = R(6420\text{A})/R(4400\text{A})$ is calculated and a histogram of the distribution of this factor is plotted in Fig. 2.25. The two distinct peaks in this figure show that two classes of Kuiper belt objects can be identified, one red with $f = 2$, the other grey with $f = 1.2$. Asteroids and 'normal' comets have f in the range 1 to 1.26 (Wickramasinghe and Hoyle, 1998; Wickramasinghe, Lloyd and Wickramasinghe, J.T., 2002). A similar conclusion follows from Jewitt's study (2005) of a larger set of Kuiper Belt objects. Significantly higher values of the redness factor f have a ready explanation in terms of fluorescence of biological-type pigments generated by solar UV processing of the surfaces of Kuiper Belt objects.

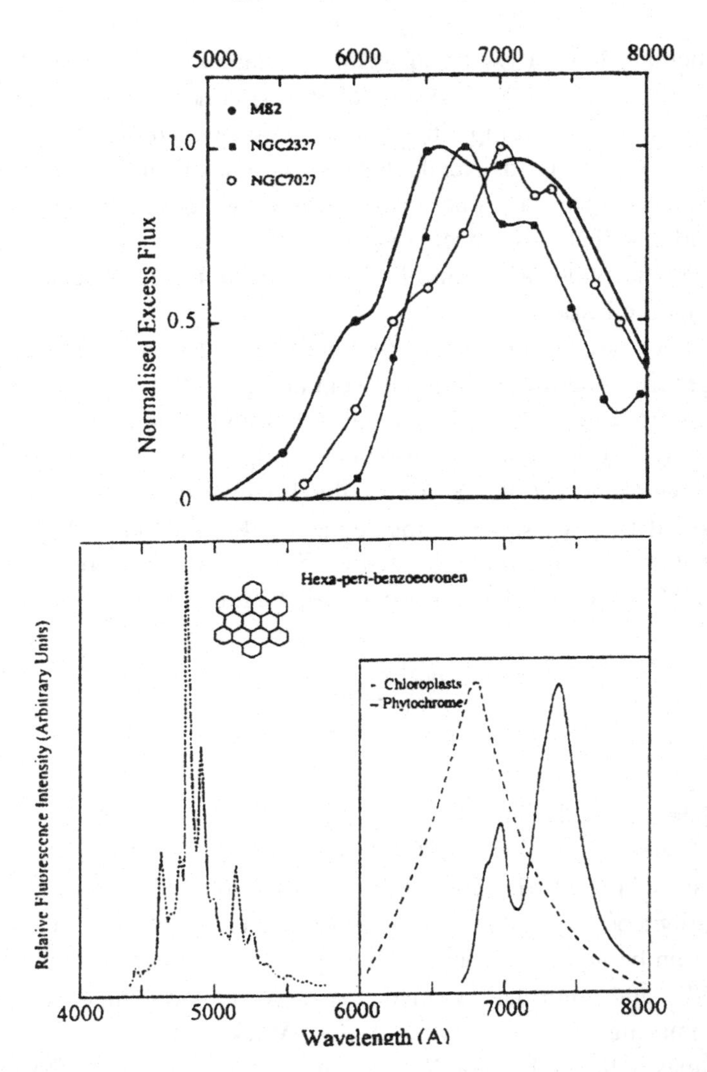

Fig. 2.24 The points in the top panel show normalised excess flux over scattering continua from data of Furton and Witt (1992) and Perrin *et al.* (1995). The bottom right panel shows relative fluorescence intensity of spinach chloroplasts at a temperature of 77 K. The dashed curve is the relative fluorescence spectrum of phytochrome. The bottom left panel is the fluorescence spectrum of hexa-peri-benzocoronene.

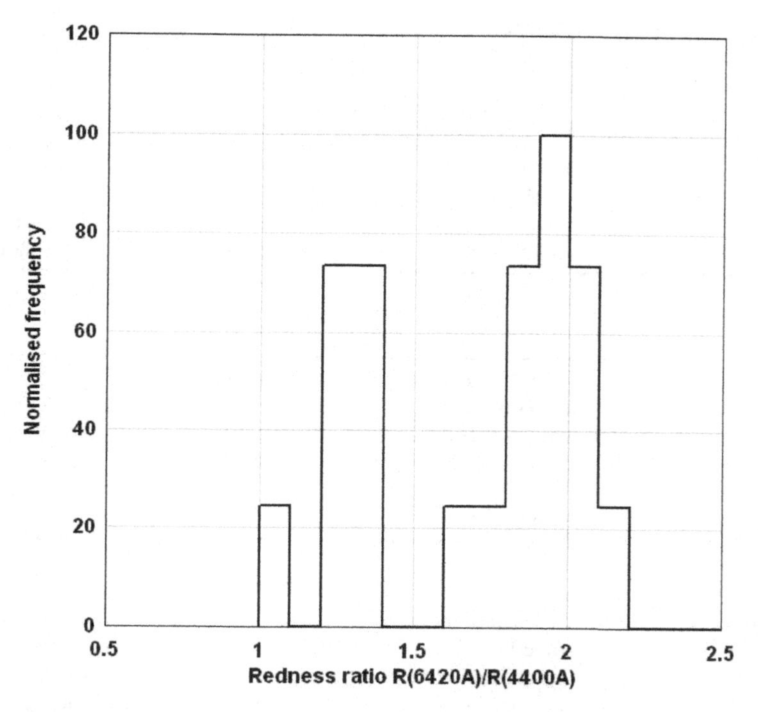

Fig. 2.25 The distribution of the reflectivity ratio in Kuiper Belt Objects and Centaurs, from data given by Tegler and Romanishin (2000).

2.10 The Origin of Organic Molecules in Space

Since the late 1960s interstellar clouds have been discovered to contain a large number of molecular species. The total number of molecules discovered thus far in the ISM total over 150 but this tally is dependent on the limitations of observational techniques involving radio, mm and IR observations. Molecules that are found to be widespread include water, methane, ammonia, formaldehyde, alcohol, glycolaldehyde and hydrogen cyanide, all of which could have a biological significance. A selection of the larger organic molecules detected to date are listed in Table 2.4:

Table 2.4 Interstellar molecules with more than 7 atoms.

	Molecules with 7 atoms		
CH_3CCH	methyl acetylene	CH_3CHO	acetaldehyde
CH_3NH_2	methylamine	CH_2CHCN	vinyl cyanide
HC_5N	cyanobutadiyne	C_6H	1,3,5–hexatriynl
c–C_2H_4O	ethylene oxide	CH_2CHOH	vinyl alcohol
	Molecules with 8 atoms		
CH_3COOH	acetic acid	CH_3OCHO	methyl formate
CH_3C_3N	cyanomethylacetylene	CH_2 (OH)CHO	glycolaldehyde
H_2C_6	hexapentaenylidene	HC_6H	triacetylene
C_2H_6	ethane	C_7H	2,4,6–heptatriynylidyne
CH_2CHCHO	propenal		
	Molecules with 9 atoms		
$(CH_3)_2O$	dimethyl ether	CH_3CH_2OH	ethanol
CH_3CH_2CN	ethyl cyanide	CH_3C_4H	methylbutadiyne
HC_7N	cyanohexatriyne	C_8H	1,3,5,7–octateraynyl
	Molecules with 10 atoms		
$(CH_3)_2CO$	acetone	$HOCH_2CH_2OH$	ethylene glycol
CH_3CH_2CHO	propanal	CH_3C_5N	methyl–cyanodiacetylene
	Molecules with 11 atoms		
HC_9N	cyanooctatetrayne		
	Molecules with 12 atoms		
C_6H_6	benzene		
	Molecules with 13 atoms		
$HC_{11}N$	cyanodecapentayne		

Organic molecules (molecules containing carbon) make up some 90% of all the gas phase interstellar molecules. Many of these are located in hot dense gas inside star-forming regions such as the Sagittarius B2 clouds near the Galactic centre. A widely-held view is that such molecules result from a combination of ion-molecule chemistry in the gas phase followed by radiation processing of mantles accreted on grains (Herbst and van Dishoeck, 2009). However, the presence of large organic molecules in cold dark clouds (e.g. Taurus Molecular Cloud complex TMC-1), where *in situ* radiation processing will be impossible, is beginning to place severe strains on current theories of molecule formation. The molecules methylcyanodiacetylene ($CH_3(CC)_2CN$) and methyltriacetylene (CH_3C_6H) recently discovered in TMC-1 could be examples that pose a problem for abiotic theories.

Recently astronomers working at the Robert C. Boyd radio telescope, which has a greatly enhanced wavelength, coverage and sensitivity compared with earlier instruments, have discovered ~1000 new spectral lines. Of these some 240 have been attributed to complex organic molecules but these have still to be identified. Many appear to be present in cold dark clouds where radiation processing will be inefficient.

We have already seen that observations of mid-infrared absorption bands and the 2175A ultraviolet extinction feature provide evidence of aromatic molecules (PAHs). Direct mass spectroscopy of interstellar dust in the *Stardust* mission (Krueger *et al.*, 2004) has also shown the presence of cross-linked heteroaromatic structures in the degradation products of impacting interstellar grains. Evidence of fragments with atomic mass unit AMU > 2000 consistent with pyrrole, furan substructures and quinones were found (Fig. 2.26). The fractured components of cell walls are arguably the only types of molecular structures that would survive impacts at speeds of ~30 km/sec.

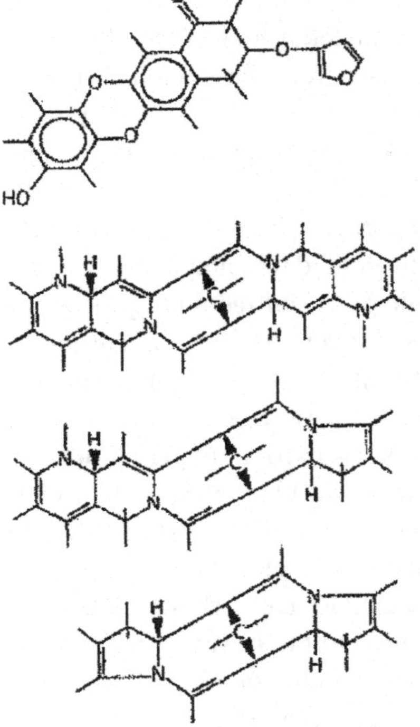

Fig. 2.26 Functional groups in the break-up fragments of impacting interstellar dust grains, inferred by Krueger and Kissel (2000) from mass spectroscopy.

2.11 Direct Analysis of Comet Dust

The first *in situ* analysis of comet dust was carried out in 1986 for dust particles from comet Halley as they impacted a mass spectrometer on board the *Giotto* spacecraft at speeds of ~30 km/s (Kissel and Krueger, 1987). By analysing the masses of break-up products, a large fraction of the grains were found to be organic ('CHON' particles) with compositions that included long-chain hydrocarbons and nitrogenated polyaromatic molecules. These molecular structures are fully consistent with the break-up of bacterial cells.

The claim originally made that bacterial particles are ruled out because the break-up material did not show evidence of the biologically important element phosphorus is flawed. Molecular ion-mass spectra of Kissel and Krueger (1987) could be interpreted as plausible combinations of P with other elements. The break-up of phosphorus groups (as in DNA) could lead to possible mass peaks corresponding to PO_3^+ (79), PO_2^+ (63) or PO^+ (47) rather than P^+ (31). That such evidence does indeed exist was shown in an analysis by one of the present authors (Wickramasinghe, 1993). The claim that mass peaks corresponding to two other biologically significant elements Na and K are too low for biology is also open to challenge. Although freeze-dried cultures of vegetative bacteria may be ~100 times richer in Na and K, this is not the case for nutrient starved bacteria nor for spores (Wickramasinghe, 1993).

More recent spaceprobe studies of comets have not resolved the question whether or not a fraction of the organic molecules in comet dust is biogenic. The question still remains wide-open, although the conventional view is that any organic molecules found in comets must of necessity be of abiotic or prebiotic origin.

The *Stardust* mission to comet Wild 2 was launched in 1999 and the experiments it carried were planned over the preceding decade at a time when cometary microorganisms were dismissed as an outright impossibility. The protocol used for particle collection would not have permitted the survival of intact cells.

In January 2004 the *Stardust* spacecraft sped through the tail of comet Wild 2 and capturing dust particles in blocks of aerogel. Each particle impacted the aerogel at an initial speed of ~6km/s excavating a track along which break-up debris was deposited. Along the tracks a wide range of organic molecules including the amino acid glycine was discovered. The detection of hetero-aromatic molecules rich in N and O could be a tell-tale sign of degraded biomaterial, biology being particularly rich in such structures. The aerogel also showed evidence of pre-solar grains including a class of highly refractory minerals which probably condensed in supernovae explosions.

The *Deep Impact* mission that reached comet Tempel 1 on 4 July 2004 involved a high-speed impactor that crashed onto the comet's surface. The crust of the comet was thereby ruptured releasing sub-surface cometary material that was analysed *in situ*. The *Deep Impact* mission found evidence of clay particles in comets for the first time, together with water and a range of organics. The implication of these findings will be discussed more fully in Chapter 7.

2.12 Capture of Comet Dust in the Stratosphere

In both the *Giotto* and *Stardust* missions cometary particles were examined and studied *after* they had endured high-speed impacts. Survival of fragile organic structures (e.g. bacteria) would have been virtually impossible in such conditions.

An obvious place to find fragile particles from comets could be the Earth's upper atmosphere. Cometary meteoroids and dust particles are known to enter the atmosphere at a more or less steady rate. Although much of the incoming material burns up as meteors, a significant fraction must survive. Micron-sized organic grains, arriving as clumps and dispersing in the high stratosphere, would be slowed down gently and would not be destructively heated. The atmosphere could thus serve as an ideal collector of organic cometary dust.

Techniques for stratospheric collection of cometary dust must of necessity involve procedures for either sifting out terrestrial contamination or for excluding contamination altogether.

Stratospheric dust collections have been carried out from as early as the mid-1960s (Gregory and Monteith, 1967). Balloons and rockets reaching heights well above 50 km were deployed and consistently brought back algae, bacteria and bacterial spores. Although some of the microorganisms thus collected were claimed to exhibit unusual properties such as pigmentation and radiation resistance, their possible extraterrestrial origin remained in serious doubt at this stage. No DNA sequencing procedure was available at the time to ascertain any significant deviations there might have been from terrestrial species. Moreover the collection and laboratory techniques in the 1960s left open a high chance of contamination.

The modern pioneer of cometary dust collection is unquestionably D.E. Brownlee. Brownlee used U2 aircraft with 'sticky paper' mounted on an exterior structure to intercept the airflow at relative speeds of some 500 km/hr. Large volumes of the lower stratosphere at a height of about

15 km were sampled in this way with aerosol particles being deposited on the sticky paper. At heights of 10–15 km a large fraction of the dust was undoubtedly of terrestrial origin, so Brownlee had to use a variety of criteria including isotopic studies to identify and pick out particles of cometary origin.

The particle shown in Fig. 2.27 is one such cometary particle showing an exceedingly fluffy structure. It is likely that the most volatile organic (or inorganic) material originally associated with this particle was stripped out as a result of the impact speed onto the sticky paper of some 500 km/hr.

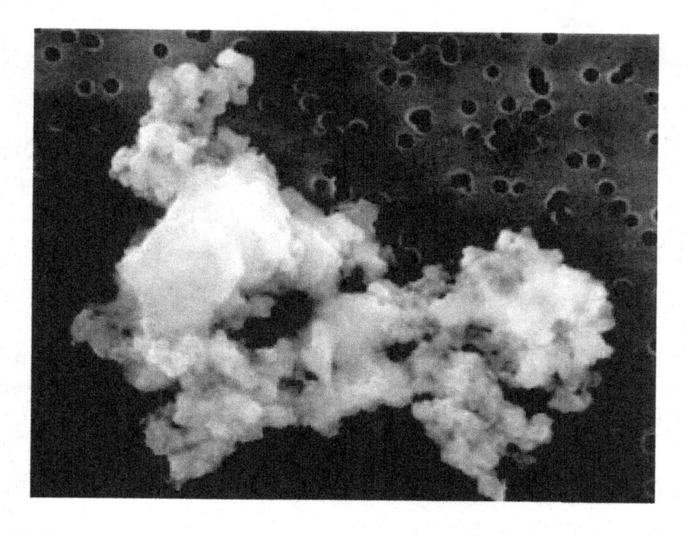

Fig. 2.27 Scanning Electron Micrograph of a Brownlee particle (Courtesy NASA).

However, studies of 8 Brownlee particles by Clemett *et al.* (1993) identified exceedingly complex organic molecules including aromatic and aliphatic hydrocarbons.

Even more significantly, a decade earlier, Bradley *et al.* (1984) described the carbonaceous structure shown on the left panel of Fig. 2.28 which was included within a Brownlee particle. Bradley *et al.* have interpreted this structure, which includes a magnetic domain, as being the result of heterogeneous catalysis. Hoyle *et al.* (1985) compared this structure with a 2 My old microbial fossil found by Hans Pflug in the

Gunflint cherts of N. Minnesota. In view of the striking similarity seen between the two images in Fig. 2.28 the most reasonable explanation might be that the particle described by Bradley *et al.* was a partially degraded iron-oxidising bacterium.

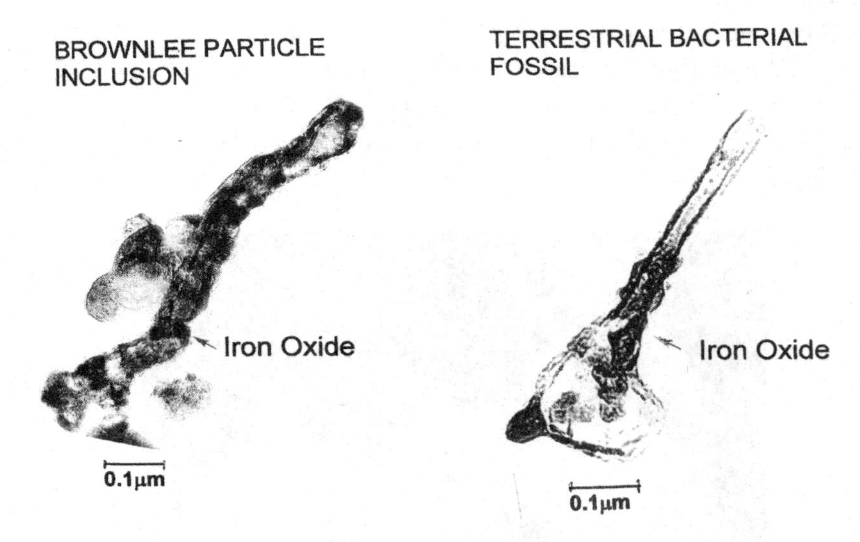

Fig. 2.28 A particle of complex organic composition collected in the stratosphere (Bradley *et al.*, 1984) compared with a fossilised iron oxidising bacterium in Precambrian sediments (Hoyle, Wickramasinghe and Pflug, 1985).

In 2001 the Indian Space Research Organisation (ISRO) launched a balloon into the stratosphere carrying devices to collect stratospheric air under aseptic conditions (Harris *et al.*, 2002). The procedure involved the use of cryogenically cooled stainless steel cylinders which were evacuated and fitted with valves that could be opened when they reached a predetermined altitude. Large quantities of stratospheric air at 41 km were thus collected and the cylinders were brought back for analysis to laboratories in Cardiff and Sheffield.

The ultra-high pressure stratospheric air contained within the cylinders was carefully released and passed through a system including millipore membrane filters. Upon such filters stratospheric aerosols were collected, extreme care being taken at every stage to avoid contamination. The particles that were collected fell into two broad classes: (a) mineral grain aggregates, very similar to Brownlee particles, but somewhat smaller; (b) fluffy carbonaceous aggregates resembling clumps of bacteria (see Fig. 2.29). Typical dimensions were about 10 μm.

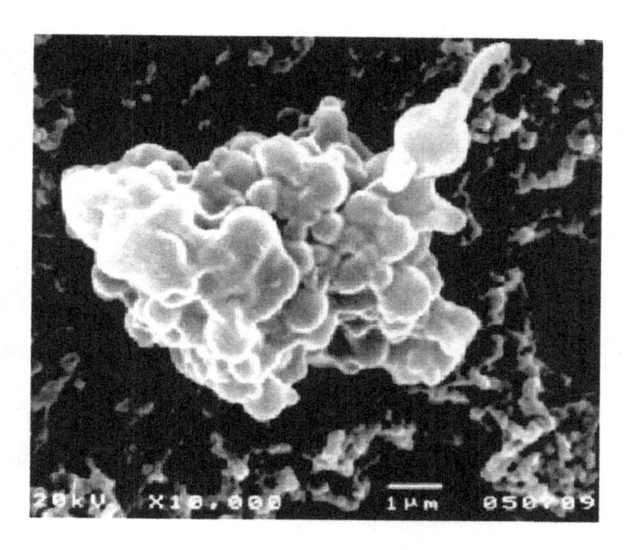

Fig. 2.29 A carbonaceous stratospheric particle from 41 km resembling a clump cocci and a rod bacterium (Courtesy: Cardiff University).

The origin of such particles is very strongly indicated, the altitude of 41 km being too high for lofting 10-micron-sized clumps of solid material from the Earth's surface. In addition to structures such as Fig. 2.29, which can be tentatively identified with degraded bacteria, the stratospheric samples also revealed evidence of similar sized bacterial clumps that could not be cultured, but were nevertheless detected unambiguously by the use of a fluorescent dye. The uptake of the dye revealed the presence of living cells in the clumps.

In a separate series of experiments a few microbial species were also cultured from stratospheric air samples by Wainwright *et al.* (2003) and there is tentative evidence that these may also have come from comets. But contamination cannot be ruled out entirely and the jury is out on all these matters.

Chapter 3

The Origin of Comets

In urban areas, where stars are barely visible, comets are almost never seen. However, away from street lights, about one comet in an average year is visible as a naked eye object. It usually appears in the sky as a faint, hazy patch, sometimes described as ghostly. It will cross the sky, moving against the background of constellations over a period of weeks to months. As it approaches the Sun, it will brighten and develop one or more tails, and these may sometimes be quite long. On rare occasions, a brilliant comet will appear, developing a smoky, dark-red tail which may sometimes seem to split the sky in two. A great comet can be seen in broad daylight. Such monsters, silently drifting across the sky, have in the past given rise to fear, consternation and thundering sermons from pulpits. It is, after all, an irruption into the established order of the heavens, and surely portends some dramatic event on Earth.

About 1500 apparitions of comets have been recorded, half of them being comets which have never returned nor, for the most part, are expected to. The recorded annual influx of these long-period comets (by convention, with orbital periods $P > 200$ years) has not changed much over the last 200 years, and corresponds to an Earth-crossing flux of about 0.8–0.9 comets/AU/yr brighter than absolute magnitude 7, comparable with Halley's Comet. Probably about 1.5 such comets pass within the orbit of Jupiter every year. The absolute magnitude of a comet nucleus is defined from:

$$m = H_{10} + 5\log_{10}\Delta + 10\log_{10} r$$

where m represents its apparent magnitude, H_{10} is the absolute magnitude, Δ is the distance from the comet to the Earth and r is the Sun-comet distance. The returns of Halley's Comet, a periodic comet with orbital period 76 years, have been recorded for about 2000 years. It is an extremely dark object (albedo 0.03±0.01) measuring 16×8×8 km. It loses volatiles from a few vents which amount to less than 10% of its surface.

3.1 The Galactic Disc

It is reasonable to assume that the origin of planets and, probably, comets is bound up with the problem of star formation generally. Stars are born out of the gas and dust of the interstellar medium. The process is wasteful, with much of the material going into the formation of stars being returned explosively to the Galactic field.

Although the interstellar medium is about 70% hydrogen, most of this element is in its ground state and difficult to observe directly.

The proton and electron in a neutral hydrogen atom are either aligned or anti-aligned, with spontaneous transitions taking place between these two states. During such a transition, 21 cm radiation is emitted and the detection of this radiation is an important tool in mapping out the structure and velocity of the interstellar medium. At very low temperatures, hydrogen occurs in predominantly molecular form. Molecular hydrogen is much more difficult to detect, but radio astronomers map out its presence in cold molecular clouds and nebular masses using carbon monoxide as a proxy, with $H_2/CO \sim 10^5$.

Several types of nebulae exist in interstellar realms. Diffuse HI clouds of atomic hydrogen are found, with temperatures typically 30–80 K and masses typically up to 100 $M_\odot$. The number density of hydrogen atoms in these nebulae is $n_H \sim$ 10–1000 cm^{-3}. They are not associated with star formation. Cold, extremely dense nebulae known as Bok globules show up in silhouette as small dark regions against a background of luminous nebulae; characteristically they are of sub-parsec dimensions, have masses typically a few hundred solar masses, and are extremely cold

($T \sim 10$ K) and dense ($n_H \sim 10^4$ cm^{-3}). They are often the sites of star formation. Such globules are extremely dusty, with dust densities about two orders of magnitude greater than those of the interstellar medium as a whole.

However the main star formation factories are molecular clouds, particularly giant molecular clouds. About half the mass of the interstellar medium is in the form of molecular hydrogen, and most of this, perhaps $\sim 2\times10^9\ M_\odot$ in total, resides in the GMCs. Supersonic turbulence in the interstellar medium appears to play a crucial role in the formation of molecular clouds and molecular hydrogen. A GMC may have radius 20 parsec and comprise several hundred thousand solar masses of material at a temperature $T \sim 10$–20 K with mean $n_H \sim 300$ cm^{-3}. It is typically an aggregate of half a dozen smaller nebulae which are themselves subdivided and so on, with a differential mass distribution $n(M)\,dM \propto M^{-1.6}$ extending down to at least ~0.1 parsec and a few solar masses. A GMC is largely empty space, with a filling factor typically 0.05.

Far infrared and X-ray observations reveal that, hidden behind the opaque exterior of a GMC, hundreds or thousands of stars may be in the process of forming at any one time. Probably about 80% of current star formation is taking place within these massive clouds. Stars do not form in isolation, but rather in clusters. Star formation is a self-limiting and inefficient process, since it involves the creation of supernovae and hot B stars with strong winds which disperse the nebulae. Thus molecular clouds probably have lifetimes only about 10 million years. Otherwise, there would be little nebular mass in the Galaxy.

The GMCs are preferentially concentrated along the spiral arms of the Galaxy, often occurring in chains or filaments, whereas less massive nebulae are distributed more uniformly throughout the Galactic plane. Molecular clouds in general comprise a thin system in the plane, with half-thickness 60–70 parsecs. The system of diffuse HI nebulae form a somewhat fatter system; it looks like a bubble bath, made up of intersecting bubbles of gas driven by stellar explosions. In all, about a third of the mass of the Galactic disc is in nebular form.

3.2 The Formation of Stars

The tendency of a discrete nebula to collapse under its own weight is opposed by the kinetic energy of its components, in particular the thermal energy of the gas which comprises it. If we neglect magnetic fields and bulk motion, and consider a uniform, homogenous cloud, we could imagine perturbations in the form of waves of length λ being induced by collisions with other nebulae, the passage of supernova shockwaves and the like. Below a critical wavelength λ_c the perturbation is an ordinary soundwave; above it, the perturbation becomes self-gravitating and collapses under its own weight. The condition for marginal stability is provided by the virial theorem

$$2K + \Omega = 0 \tag{3.1}$$

where K is the thermal energy of the perturbation — or we can equally consider an isolated nebula — and Ω is its gravitational potential energy. The kinetic energy of the atoms and molecules is given by

$$K = \frac{3}{2} NkT = \frac{3MkT}{\mu} \tag{3.2}$$

Here N is the number of molecules, M is the mass of the assembly, μ the molecular weight (2 in the case of molecular hydrogen), T the temperature and k is Boltzmann's constant.

The gravitational potential energy is given by

$$\Omega = \int_V \boldsymbol{r} \cdot \boldsymbol{g} \rho \, dV \tag{3.3}$$

where $\boldsymbol{g}$ represents the local gravitational acceleration within the perturbation, and the gravitational potential $\boldsymbol{r.g}$ of an element of mass ρdV is integrated over the whole volume V. In the case of a uniform spherical nebula of mass M and radius R,

$$\Omega = -\frac{3GM^2}{5R} \tag{3.4}$$

G the gravitational constant. Putting these equations together we readily find the Jeans length

$$R_J = \sqrt{\frac{15kT}{4\pi\rho G\mu}} \tag{3.5}$$

and the corresponding Jeans mass

$$M_J = \left(\frac{5kT}{G\mu m_H}\right)^{3/2} \left(\frac{3}{4\pi\rho}\right)^{1/2} \tag{3.6}$$

named after James Jeans, who derived the equation in 1902. For a uniform sphere this reduces to

$$M_J \sim 35\sqrt{\frac{T^3}{n_H}}\, M_\odot$$

The corresponding critical radius is

$$R_J \sim 6.5\sqrt{\frac{T^3}{n_H}} \text{ parsec}$$

In the case of diffuse HI clouds the critical mass is typically over 1000 $M_\odot$. If we take the core of a molecular cloud to have $n_H \sim 10^4$ atoms per cubic centimetre and a temperature 10 K, then with $\mu = 2$ for molecular hydrogen, we find an initial Jeans mass $M_J \sim 35\, M_\odot$.

In the early stages, radiation is lost efficiently and the collapse is isothermal. Collapse occurs in freefall on a timescale

$$t \sim 1/(G\rho)^{1/2}$$

independently of the size of the initial perturbation. Cloud collapse during this phase is quite rapid, occurring on a characteristic timescale of about 10,000–100,000 years, since the cooling timescale due to dust is much shorter than the freefall collapse time. During this phase the Jeans mass declines, initial inhomogeneities in the mass are amplified and subregions of the collapse in cloud detach themselves and shrink independently. It seems that cloud-cloud collisions at small scales are adequate to induce star formation, and can reproduce the mass distribution of proto-stellar cores within molecular clouds.

Also during this stage, since angular momentum has to be shed, a disc of material a few hundred astronomical units in radius develops around the collapsing protostar, along with bipolar flows. Magnetic fields, which rapidly strengthen as a protostar collapses, must play an important role in transferring angular momentum from the growing protostar to the material around it.

This early collapse and hierarchic fragmentation will proceed to the point where the cloud becomes optically thick due to the concentration of dust, a situation which arises when $n_H \sim 10^9$ cm^{-3}. Fragmentation stops at about a solar mass. From this point, with heat less able to escape, the thermodynamic situation is more nearly adiabatic and the protostar's rate of contraction is determined by the loss of heat from its surface. High optical thickness is reached first in the core, which becomes hydrostatic while surrounding gas is still in freefall. With ionisation of the outer layers, and the resulting increase in opacity, radiative transfer becomes insufficient for losing the heat being generated by gravitational contraction, and heat loss from the outer layers of the star becomes governed by convection. This stage cannot generally be observed because of the obscuration by the surrounding dusty gas. But when the accretion has more or less finished, the protostar is now on a Hayashi track, slowly collapsing from an high initial luminosity while its effective temperature stays much the same. In the case of the Sun, the Hayashi track was reached when the protostar had a surface temperature ~3000 K and a luminosity perhaps 1000 $L_\odot$.

The collapse time is then given by the Kelvin–Helmholtz timescale

$$t_{KH} \sim 50\, M^{2} L^{-1} R^{-1}\, My$$

with (*M, L, R*) in solar units. When the temperature in the core reaches about 1,000,000 K, deuterium burning begins, followed by the proton proton chain reaction, speeded up by the ^{3}He created in the earlier deuterium burning. The protostar has now reached the zero-age main sequence and is a fully fledged star.

3.3 Planet Formation

New observational techniques such as long-baseline interferometry at sub-millimetre wavelengths, and speckle imaging in the near infrared, now allow astronomers to investigate nearby protoplanetary discs at sub-arcsecond resolution. One result to emerge is that circumstellar discs are larger than had been expected: the accretion discs observed around low-mass stars are generally 100 or more AU in radius, several times the size of our planetary system. Structure — possibly preplanetary condensations — has been detected, at infrared and submillimetre wavelengths, in several debris discs at distances up to several hundred AU from the central star, as in AB Aurigae (Grady *et al.*, 1999). Strong winds from the young star are assumed to eject gas from at least the inner regions of the disc, leaving dust and larger bodies up to 100 km across in the remnant. It has been suggested that the growth of a large planet in the disc would generate a spiral instability in the dust disc and that this instability would cause planetesimals and dust further in to aggregate into small planets. Thus Jupiters, it is proposed, may generate Earths. There is evidence that disc accretion may continue for a few million years after the formation of a star, during which time planet formation is ongoing (Jayawardhana *et al.*, 2006).

Few observational data are available for accretion discs around more massive stars, which are rare and are therefore seen at larger distances. The formation of such stars may proceed by monolithic collapse of isolated cores, by competitive accretion of infalling clumps of matter within a protocluster, or even perhaps through stellar collisions in very dense starforming regions (Bonnell and Bate, 2006). The lifetimes of such stars are very short before they explode, characteristically 10^7 years for a 10 $M_\odot$ B star, which is much too short a timescale for life to evolve significantly on any habitable planets around such stars. At the other extreme, stars with masses less than 0.08 $M_\odot$ lack sufficient mass for nuclear burning, and again life in habitats around such brown or black dwarfs is unlikely to survive beyond the Kelvin–Helmholtz contraction time of about 50 million years. The most favourable planetary habitats for long-lived evolution would seem to be found in planets in near-circular orbits around long-lived stars of moderate to low mass. As it happens most of the mass of the Galaxy resides in such stars, the initial mass function peaking at about 0.3 $M_\odot$.

A consideration is that most, if not all, T Tauri stars are binary or multiple systems, and this may limit the number of suitable environments available for the evolution of life. The binary frequency decreases to about 50% for the nearby solar-type stars, possibly because of disruption by stars or molecular clumps in the proto-cluster. Planetary systems appear to be quite common in binary and triple-star systems (Raghaven *et al.*, 2006).

The formation of planets out of an accretion disc is not well understood, as many processes are involved. It is uncertain whether gas giants, for example, grow by accretion of gas around large rocky cores or by gravitational instability of the gaseous disc. For smaller planets, the growth of multi-kilometre planetesimals via the accretion of dust to pebbles to boulders requires an efficiency of coagulation at all scales which may be unrealistically high. The settling of dust towards midplane in the midst of thicker gas generates instabilities at the dust/gas interface which may not allow the necessary dust concentrations to develop (Barranco, 2009). Collective effects due to the mutual gravity of many accreting boulders may also be disrupted by such turbulence.

The formation of the terrestrial planets at least seems to have been rapid, since basaltic lava flows on the surface of the asteroid Vesta were formed within 3 million years of the origin of the solar system. This is consistent with observations of the discs around several young stellar objects which indicate that planetary systems are formed within a few million years of the arrival of the stars on the main sequence (Yin *et al.*,

2002). What does seem clear is that the process of planetary formation was violent, at least in our solar system. Evidence for this can be seen in the fact that the rotation axis of the Sun is tilted 7° to the plane of the ecliptic, that the rotation axis of Neptune is tilted 98° from its orbital plane, and that the Moon appears to have formed by material torn from the Earth by the collision of a Mars-sized body.

3.4 The Formation of Comets

The picture of planet formation which has emerged in recent years is backed up by numerous examples of accretion discs which reveal evidence of planets embedded within them. It seems reasonable that we should be able to fit the origin of comets into this scheme.

One consideration is that the Sun was in all probability born in a star cluster such as that which we see in a number of molecular clouds, for example the Trapezium nebula in Orion. But in that case the Sun is unlikely to have been the first to emerge in the cluster. The more massive the protostar, the more rapid its collapse down the Hayashi track onto the main sequence. A 15 $M_{\odot}$ star, for example, will reach the main sequence in about 50,000 years, a 3 $M_{\odot}$ star in two or 3 million years, about 5% of the collapse time of the protosun. Thus the early Sun was probably born in an environment where comets had already formed in large numbers.

3.4.1 The structure of the comet population

Most long-period comets have periods measured in millions of years and arrive from distant locations with almost zero gravitational energy with respect to the Sun. In 1932 Öpik suggested that the Sun must be surrounded by a cloud of such comets, ejected there from an inner region by passing stars. However he seems to have overlooked that stellar perturbations would also throw some comets back into the planetary system. This latter proposition was made by Oort in 1950. He thought the radius of the cloud was about 200,000 AU. Since, at that distance, the planetary system is a tiny speck on the celestial sphere, it was a simple matter to estimate the total population of this comet cloud on an assumption of an isotropic distribution of velocity vectors of the comets. In this way Oort considered that the planetary system was surrounded by a spherical cloud of 2×10^{11} comets with inner radius 40,000 AU and outer radius 200,000 AU. With the large increase in the long-period comet dataset of the last 50 years, and a better understanding of the nature of the perturbers, a more refined estimate is that the Oort cloud comprises $\sim5\times10^{11}$ comets down to 1 km in diameter with aphelia in the range 3,000–100,000 AU. At these larger distances, the comets are only just bound to the solar system, and are perturbed not only sporadically by passing stars and nebulae, but also continuously by the Galactic gravitational field.

The short-period comets are divided into Jupiter-family ($P<20$ yr) and Halley-type comets ($10<P<200$ yr). At present about 350 Jupiter-family comets are known, as against about 50 Halley-types, although discovery is far from complete. Whereas the Halley-type comets form a near-spherical system, with about equal numbers of prograde and retrograde orbits (comet Halley itself is in a retrograde orbit), the Jupiter-family comets comprise a much flatter, prograde system, with mean inclination ~20°. They are so-called because their dynamical evolution is dominated by frequent close encounters with Jupiter: their aphelia tend to concentrate around the radius of Jupiter's orbit, while their arguments of perihelion tend to align with that of Jupiter.

Given that the Sun and Jupiter dominate the dynamics of the interplanetary bodies, we can make use of the Tisserand parameter in the classical 3 body problem to distinguish between dynamical classes. This parameter is given by

$$T_J = \frac{a_J}{a} + 2\sqrt{a\left(1-e^2\right)/a_J}\ \cos i$$

where $a_J = 5.2$ AU represents the semimajor axis of Jupiter's orbit while (a, e, i) are respectively the semimajor axis, eccentricity and inclination of a cometary orbit. The usefulness of T_J lies in the fact that it is roughly preserved during a close encounter, and may give some clue about the dynamical evolution of different classes of body. For example in the circular restricted 3 body problem, bodies with $T_J > 3$ cannot cross the orbit of Jupiter (if assumed strictly circular), remaining forever outwith or within it.

The Tisserand parameter for most observed Jupiter-family comets lies in the range $2 < T_J < 3$, whereas long-period and Halley-type comets have $T_J < 2$. There is thus a clear dynamical distinction between the classes, and given that T_J remains roughly constant during a close encounter, it has been argued that the Jupiter-family comets cannot therefore be derived by perturbations from the Oort cloud. Numerical integrations of near-Earth and Jupiter-family comets (perihelion distances <1.3 AU) reveal that they will be ejected from the solar system, fall into the Sun or collide with a planet, with a dynamical half-life varying from 150,000 to 450,000 years (Fernández *et al.*, 2002). The system must therefore be replenished, on this timescale, from elsewhere.

Beginning in 1992, advances in computing power, the development of high efficiency CCDs, and the application of automated search procedures on large telescopes have led to the discovery of several comet populations on the fringes of the planetary system and beyond. Only the largest bodies of the populations can be detected at such distances and our understanding of them is very incomplete. The first of these populations lies just beyond the orbit of Neptune, in a belt with inner radius 30 AU and outer radius 50 AU. Its existence was predicted by Edgeworth (1949) and Kuiper (1951). The first members of the Edgeworth–Kuiper (EK) belt were discovered in the 1990s (apart from Pluto, discovered in 1930), and it probably comprises some tens of millions of inactive comets, of which about a thousand have so far been discovered. This is a kinematically cold disc strongly populated at the 3:2 and 2:1 resonances with Neptune (39–48 AU). Its total mass may be ~0.1 that of the Earth.

For some time this population was considered to be the probable source of the Jupiter-family comets. However it turns out to be too small and too stable, although some leakage into the Jupiter-family system is bound to occur. Pluto, a member of the EK belt, has been designated as a dwarf planet since 2006 — there are several similarly sized bodies in the outer fringes of the planetary system. A conspicuous feature of the dynamical theories is that they have consistently failed to predict new observational findings. We now know that the small body population on the fringes is in fact quite complex, and includes objects which seemed to be dynamically uncoupled from the planets.

A second system comprises $1\text{–}3\times10^{10}$ so-called trans-Neptunian objects, in orbits with $28<q<35.5$ AU, that is lying between Neptune and the EK belt, with $60<a<1000$ AU. These may be erstwhile EK bodies which have been scattered by Neptune.

And there is a third population of scattered disc objects with high eccentricities and inclinations (therefore not really comprising a disc), and perihelion $q > 35$ AU putting them beyond the influence of the giant planets (Neptune has semimajor axis 30 AU). The fact of being far from planetary influences creates the problem of how they got there. Sedna, about 1600 km in diameter, has perihelion $q \sim 75$ AU and semimajor axis $a \sim 500$ AU. Eris, in an orbit with $a \sim 68$ AU, $q \sim 38$ AU and orbital inclination $i \sim 44°$, is currently the largest known object in these outer reaches. With a diameter 2600 km, it is marginally larger than Pluto at 2300 km. It is 27% more massive. The surface has a high albedo (~0 .8) implying a coating of reflective ice. It is highly likely that other such bodies will be discovered.

The sculpting and architecture of these distant populations is still being explored dynamically and observationally. There are a number of unsolved problems: for example the current EK belt is about a factor of 100 too small to have grown by *in situ* accretion; if it did, somehow 99% of the mass has vanished. Nor is it clear how the scattered disc population could have 'scattered' from the region of the planets. However significant progress has been made by modelling the Oort cloud taking account of the combined effects of planetary, stellar and Galactic perturbations (Emel'yanenko *et al.*, 2007). It turns out that in broadbrush terms the Oort cloud is the probable source of most comets which pass through the inner solar system. About 1% of Oort cloud comets entering the planetary system are perturbed by the giant planets into Halley-type orbits, and it turns out that even the Jupiter-family comets may originate from the Oort cloud: the Tisserand barrier, it turns out, can be crossed. This is not to say that all Jupiter-family comets necessarily derive from the Oort cloud. The inverse process of populating the Oort cloud from the EK belt, seems to be inadequate. At least some of the scattered disc population, dynamically uncoupled from the planets, is also accounted for by transfer from the Oort cloud.

Thus the upshot of modern dynamical studies is that the Oort cloud plays a fundamental role in determining the flux of comets through the planetary system, with the EK belt and other local sources being of secondary importance.

One consequence of this finding is that a record of past Galactic perturbations may be preserved in terrestrial impact craters. Another is that, in trying to determine the origin of comets, we should focus on the origin of the Oort cloud. And in hunting down the conditions of formation of comets, we need to investigate their chemistry.

3.4.2 Comet chemistry

Comets contain roughly equal proportions of water-ice and dust, the dust being mostly organic as we saw in Chapter 2. Some comets are more dusty than others. As they approach the Sun comets begin to outgas. The 1986 studies of comet Halley showed clearly that outgassing is a far more complex and capricious process than simple sublimation from a heated icy surface. Comets, after several perihelion passages, could develop organic crusts that are many centimetres thick, and such crusts could be ruptured, erratically releasing gas and dust. Comets then develop comae composed of sublimated molecules and dust.

Once caught up in the solar wind a comet will develop one or more tails, one tail composed of dust driven by solar radiation pressure, and another composed of ionised molecules interacting with the charged solar wind. Water-ice makes up more than half the cometary ices, but many other molecules are present, some of which turn out to be diagnostic of the site of origin of the comet. There is considerable diversity in the relative abundances of molecules between comets. CO, for example, varies from ~30% relative to water in some comets down to less than 1%, possibly reflecting its high volatility and different comet histories, although the chemical diversity of comets in general shows no clear correlation with dynamical class or dust/gas ratio (Biver, 2002).

As a long-period comet enters the planetary system, it passes through various 'snow lines' for different cometary volatiles. Exposed hypervolatiles such as carbon monoxide and methane would sublimate beyond the orbit of Neptune; carbon dioxide and hydrogen sulphide at about the distance of Uranus; ammonia just beyond Saturn; hydrogen cyanide between Jupiter and Saturn; and water at about 2.5 AU, about the distance of the asteroid belt.

Sequential sublimation from a cometary surface comprised of volatiles and organic dust leads to the development of a low-albedo porous crust (Hoyle and Wickramasinghe, 1986). The visibility zone of a comet is generally taken as 5 AU, but a comet may not be easily observed until it begins to lose water. The prodigious output of dust and CO from comet Hale-Bopp at a heliocentric distance of 6.5 AU is suggestive of high-pressure release for liquefied sub-surface domains rather than sublimation for a sunlit patch of frozen CO (Wickramasinghe, Hoyle and Lloyd, 1996).

Disruption of molecules in the coma means that daughter species are generally observed in the visible and UV, with the parent species having to be inferred. Parent molecules can sometimes be identified from infrared and submillimetre spectroscopy, however. Over two dozen cometary molecules have so far been identified. There is a remarkable overlap with molecules identified in interstellar clouds in warm molecular cores. It seems that whatever processes were creating molecules in the interstellar medium were also creating cometary molecules and cometary-type grains

3.4.3 Cometary origin inferred

Molecules such as water, ammonia and methane may exist in different spin species, the ratios giving the temperatures at the time of formation of the molecules. Results are available for about 14 comets, about half of them being Oort cloud and most of the others Jupiter-family. They show a remarkable uniformity, namely a formation temperature of 30 K (Crovisier, 2007) with only a few degrees deviation from comet to comet. Different molecules yield the same result. This temperature is characteristic of a molecular cloud environment and does not seem to be consistent with an origin in the Jupiter–Saturn region followed by expulsion. The proportion of hypervolatiles varies from comet to comet, but there does not seem to be any correlation between the abundance of hypervolatiles and the dynamical class of comets.

Ethane and methane were detected in comparable abundances in the bright comet Hayakutake. A solar nebula in chemical equilibrium would have yielded an ethane/methane ratio $<10^{-3}$ (Mumma *et al.*, 1996). Ethane is however readily produced by photolysis on grain surfaces in the dense cores of molecular clouds.

Photo 3.1 Comet Hale–Bopp — a long period comet — reached its last perihelion on 1 April 1997. It was one of the brightest comets seen in recent times. It had a highly elongated orbit with high inclination, almost perpendicular to the ecliptic. The orbital period as it approached perihelion was 4200 years. Organic molecules and silicate particles were observed (spectroscopically) in the comet's tail (Courtesy: John Laborde).

One of the brightest comets to arrive in the 20th-century was comet Hale–Bopp in 1995. It was 7.5 astronomical units from the Sun on discovery, a record at the time. Its diameter of 60±22 km made it six times larger than Halley's Comet. It was visible for 18 months, reaching a negative magnitude for eight weeks and giving astronomers a unique opportunity for intensive study. The relative deuterium abundance D/H varied from compound to compound in the coma, which seemed to suggest an origin in an interstellar cloud environment. The temperature of formation seems to have been 25–30 K. The detection of argon, a highly volatile noble gas, along with krypton at depleted levels, brackets the past temperature history of the comet in the range 20–35 K which, with conventional solar nebula models, again gives it a formation site somewhere beyond the planetary system.

In January 2004, the *Stardust* mission collected samples from comet Wild 2 and returned them to Earth two years later. This comet had a close encounter with Jupiter in September 1974 which took it from the EK belt into its current orbit. Its residence time in the EK belt is uncertain but likely to have been very long. The flyby approached within 234 km of the comet's surface, and collected material from the coma which had been exposed to sunlight for a few hours. Over 5000 particles up to 300 μm were collected. The analysis of these particles, by over 200 groups worldwide, revealed a remarkably heterogeneous range of organics whose chemistry has been claimed to be indicative of radiation processing of ices in dense molecular clouds, and the enrichment of individual grains with ^{2}H and ^{15}N is likewise suggestive of a presolar, interstellar heritage (Sandford, 2008). But, as we pointed out in earlier Chapters, a large fraction of the organics detected in the *Stardust* experiments could arise from the break-up of putative biological particles (Coulson, 2009). This would be an option that we shall keep open in our further discussions in this book.

On the other hand, much of the mass of the comet seems to have come from material in the hot, inner regions of the solar nebula. These indications include the presence of refractory grains of solar system origin, and grains with helium and neon isotope ratios common to both *Stardust* samples and carbonaceous material in meteorites. The returned grains resembled meteoritic material from the asteroid belt, composed mostly of inner solar system minerals (Ishii *et al.*, 2008). Even more extraordinary was the identification of a particle whose microstructure resembled that of lava from a small, molten, rapidly cooling body (Leroux *et al.*, 2008).

Grains rich in iron oxide have been recovered and interpreted as evidence that liquid water was present at some stage in the history of Wild 2 (Bridges *et al.*, 2008).

3.4.4 Other ideas about comet origins

Ideas about the origin and nature of comets are older than Christianity (Bailey *et al.*, 1990). The main debate in the 19th and early 20th centuries was whether comets have an interstellar origin, or are indigenous to the solar system. The idea of an interstellar origin was revived by Lyttleton (1948). This arose out of work by Bondi, Lyttleton and Hoyle in which the accretion of matter onto stars while moving through nebulae was examined theoretically.

In Lyttleton's scheme the Sun, moving through a cloud of dust, generates a wake of particles through gravitational focusing along the axis downwind of the Sun's motion. Some fraction of the hyperbolic energy would be removed by the mutual collisions of the particles concentrated along the axis. Within a critical heliocentric distance r_0 enough energy is lost for the material to be captured and so fall in towards the Sun. This distance turns out to be $r_0 \sim GM_\odot/V^2$ which, for an encounter speed $V = 10$ km/s, is only about 9 AU. An encounter speed of 1 km/s would be required to yield comets with aphelia $\sim 10^3$ AU, which is still well short of the Oort cloud. To account for the formation of comets in this stream, Lyttleton assumed that self-gravity would break it into individual proto-comets, like water from a tap breaking into droplets. However, the maximum mass of the condensations turns out to be $m_c \sim 10^{20}/V^9$ g, V in km/s (Bailey *et al.*, 1990). Even at an encounter speed of 1 km/s this is orders of magnitude short of the masses of comets such as Sarabat, and in any case it is extremely unlikely that the Sun has ever encountered a nebula at such a low speed. Encounters are most frequent with giant molecular clouds which, with surface escape velocities ~20 km/s, are sufficiently massive to ensure that the low speeds required for Lyttleton accretion are never attained. There is in any case no molecular cloud in the antapex direction with the low speed required: there is little neutral or molecular hydrogen within 100 parsecs of the Sun, and the presence of soft X-ray emission of local origin indicates that we are currently inside a bubble of hot, ionised gas (Frisch, 2007).

The carbon isotope ratio $^{12}C/^{13}C$ in the interstellar medium has declined over the age of the Galaxy due to enrichment by supernovae. The solar system value of 89±1 presumably reflects that in our part of the

Galaxy 4.5 Gy ago, while the contemporaneous interstellar environment has $^{12}C/^{13}C = 77\pm10$. There is clearly a gap between the solar system and current interstellar values for this isotope ratio. The ratio determined for nine comets, eight of them from the Oort cloud, is 90±4, clearly giving them a solar system provenance.

Another possibility which is occasionally raised is that comets are fragments of an exploded planet. This idea has its roots in the 18th century, when Olbers thought that the asteroid belt might have been so caused, and was proposed by Oort in 1950 to account for the comet cloud which now bears his name. It was developed in modern times by van Flandern (1978), who proposed that at least half a dozen planets have exploded in the course of solar system history, the most recent occurring 5.5 million years ago at a distance of 2.8 astronomical units from the Sun. On this interpretation, the current long period comets are making their first return to the planetary system. The energy per gram required to disrupt a planet of mass M and radius R is $E \sim GM/R$. If Mars-sized, the body would have to be made of something with the explosive energy of dynamite (7,500 joules/g specific energy). It is not clear how this could be contrived, and van Flandern postulated 'new physics' to achieve the explosions. There was no explanation of why bodies ranging from asteroids to gas giants have not succumbed to this new physics, and the *ad hoc* nature of the exploded planet hypothesis has precluded its being taken seriously, in spite of its respectable pedigree. A modification of this idea by Drobyshevski (1978, 2008) is that the icy crusts of the satellites of the giant planets explode from time to time consequent on the buildup of electrostatic charge from the solar wind. Both this proposal and the exploded planet hypothesis are inconsistent with the extremely low formation temperatures deduced for comets, which are more akin to those appropriate to molecular clouds.

In summary, an origin of comets by the aggregation of icy grains in the primordial solar nebula or regions beyond, but still within the nascent star cluster, seems most likely based on all the current evidence.

Chapter 4

Comets in the Galactic Environment

Although, as we have seen, the concept of panspermia is older than Christianity, it is only in recent times that the subject has moved from speculation to serious scientific discussion. The idea that viable microorganisms may be transferred between suitable habitats in the solar system was revived following the discovery of relatively unshocked Martian meteorites on Earth, in particular the Antarctic meteorite ALH 84001. It was also found that the meteorite had never been subjected to a temperature much more than 40°C. However, what really raised the profile of interplanetary panspermia was the claim made for the presence of microfossils inside this meteorite. This claim remains controversial.

4.1 The Mechanism of Lithopanspermia

When an asteroid or comet strikes a planet, most of the ejecta are shocked and heated to temperatures which are lethal to lifeforms. However a small percentage of material, a few metres below the surface of the target layer, is ejected without much shocking because of the accidental cancellation of shock waves. Some of these ejecta reach escape velocity, penetrating the atmosphere quickly (if the atmosphere has not been thrust aside) and may reach interplanetary space with little internal heating and with relatively little damage to any microorganisms within them.

The identification of dynamical highways connecting the Earth and Mars, with transfer times generally $\sim 10^4$–10^7 yr (but which could be as short as ~ 1 yr), gave further credence to the idea that life may have moved between these planets, in the form of microorganisms trapped inside boulders — a process which has been called 'lithopanspermia'. Probably millions of such transfers have taken place. Mars cooled more quickly than the Earth and might have become habitable sooner (Davies, 2003). On this model, life may have migrated widely throughout the solar system, occupying every habitable niche such as probable subsurface oceans on Enceladus and Titan, and may have become embedded in the top few metres of icy bodies in the scattered disc population (Wallis and Wickramasinghe, 2004).

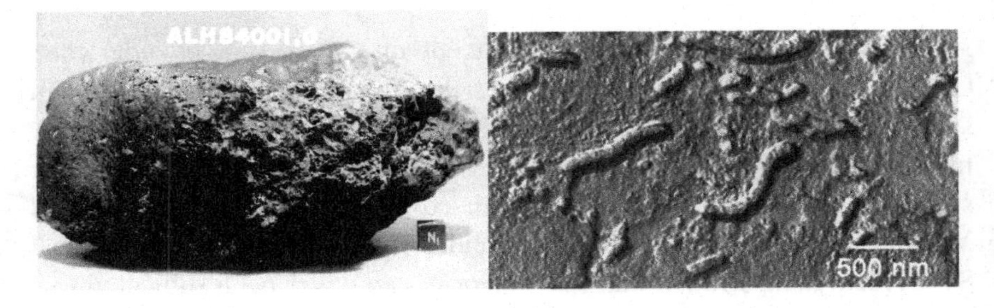

Photo 4.1 Mars meteorite ALH84001 weighing 1.9 kg, and about 18 cm long. The cube marked N is 1 cm in size.

4.1.1 Transferring boulders between planetary systems

The transfer of life-bearing boulders from the solar system to exoplanets is a much more formidable proposition. Melosh (2003) followed the fate of boulders ejected from the Earth numerically and found that, after 50 million years, ~15% of them had been expelled from the solar system, the rest taking longer.

The known cratering rate on Mars, coupled with theoretical estimates from impact mechanics, has been used to estimate the rate at which boulders over 0.2 m across have been ejected from the planet over the last 4 Gy: this turns out to be ~1000 per annum, on average (Mileikowsky *et al.*, 2000), about 10^2 times higher than for boulders thrown from the Earth. Gladman *et al.* (2000) found that about 3–10% of Martian boulders are eventually thrown into interstellar space.

If a boulder, wandering through interstellar space, encounters another planetary system, it will enter the system at hyperbolic speed. To have any chance of being captured, that system must contain at least one giant planet capable of significantly deflecting the boulder — extracting enough kinetic energy to put it into a bound orbit. If we assume almost arbitrarily that one star per cubic parsec in the Galactic field contains a giant exo-planet, then a boulder with a drift speed 30 km/sec will pass within 1 AU of a giant planet every 3 Gy. The probability that such an encounter would lead to capture into the exoplanetary system is likely to be in the range 10^{-4}–10^{-6}, and the further probability that the boulder, having been captured, will then end up falling onto a biofriendly planet in the system — should such exist — is probably of the same order. Given these unlikelihoods, it may be questioned whether a boulder from Earth has ever landed on the surface of an exoplanet. And given the enormous timescales involved in the transfer, it is to be expected that any life within the boulder would long since have been extinguished.

The converse, of a potentially life-carrying boulder from any planetary system in the Galaxy impacting on the Earth during the latter's first 700 million years, has probability only $\sim 10^{-8}$ according to Valtonen *et al.* (2009). However, these authors point out that if the solar system was born in a star cluster, $10^{2\pm2}$ boulders could have landed on the Earth from neighbouring stars; thus one life-bearing planetary system could have fertilised any other biofriendly ones which happened to be present in the cluster. Adams and Spergel (2005) find that, at best, about one or two lithopanspermia events would happen within a cluster before its dispersal.

In summary, these arguments imply that — apart from very limited transfer within a young star cluster — simply tossing boulders out of planetary systems in the Galactic disc is not a viable mechanism for interstellar panspermia. If it were the only mechanism available, then life-bearing planets in the Galaxy would develop in biological isolation from each other (of course, panspermia could be artificially spread by some advanced intelligence).

4.1.2 Erosion of ejected boulders

Two factors missing from such discussions are first, the fate of a boulder while it is orbiting within the planetary system, and second, the fact that from time to time the solar system must encounter molecular clouds within which stars are forming. We discuss the first of these matters here.

On being ejected from the Earth, a boulder becomes part of the zodiacal cloud and is subject to erosion by impacting dust particles. 'Dust' conventionally refers to particles with radii less than 100 μm. About 95% of the zodiacal light is due to such particles. The size also corresponds to a fairly sharp transition at 1 AU between radiation-dominated and collision-dominated dynamics: below this, the lifetimes of particles are limited by in-spiralling due to a radiative drag (the Poynting-Robertson effect); above it, their lifetimes are set by mutual collisions. Over timescales of order a million years or less, a Chiron-sized object will be thrown into a short-period orbit in the inner planetary system. Numerical simulations allowing for both Poynting-Robertson drag and collisional disintegration indicate that, during the disintegration of such an object, the mass of the zodiacal cloud will be enhanced for some millennia by two or three powers of ten above its current value of 10^{20} g. Over a million year period, say, there are likely to be several epochs when ejected boulders are subject to rapid destruction by erosion. The bulk of the $\approx 10^8$ yr lifetime of a boulder thrown from Earth is spent in the inner planetary system, ejection into interstellar space being quite rapid once it reaches the Jupiter/Saturn region. The question is whether the boulder would survive these episodic surges in the zodiacal cloud mass.

Consider a boulder of bulk density ρ and radius r immersed in zodiacal cloud particles with space density ρ_z and which strikes the boulder at V km/sec. Over a time Δt the boulder loses a mass Δm given by

$$\Delta m = -\pi \rho_Z V \Gamma \Delta t$$

and its radius decreases by Δr obtained from

$$\Delta m = 4\pi \rho r^2 \Delta r$$

Here $\Gamma = m_e/m_z$, where m_e is the mass of material excavated by a colliding particle of mass m_z. Thus

$$\Delta r = -\frac{1}{4}\left(\rho_Z/\rho\right) V\Gamma\Delta t$$

and so the radius of the boulder $r = r_0 - kt$ decreases linearly until it disappears in a time

$$\tau = 4r\rho/\left(V\Gamma\rho_z\right)$$

The excavation factor Γ is found experimentally to vary as V^2 and has a value, for medium strength rock with impact speed 10 km/sec, of $\Gamma \sim 5\times10^4$ (Grün *et al.*, 1985). In the present day zodiacal cloud, a 1 m boulder would be destroyed by erosion within 20,000–250,000 years, depending on the uncertain mass of millimetre sized meteoroids. However in the presence of a large disintegrating comet, the cloud mass is enhanced by a factor of 10–100, while the particle velocities, due to the higher eccentricities involved, are enhanced by at least 50%. The erosion rate ΓV^2 varies as the cube of the velocity and so the boulder erosion time reduces to a few centuries or millennia, substantially less than their residence time in the inner planetary system.

Once the boulders have been reduced to 'dust' then subsequent collisions reduce the dust to β-meteoroids, for which solar radiation pressure exceeds gravitational acceleration, and the particles are accelerated out of the solar system. The velocities attained can reach some kilometres per second allowing the grains to travel some parsecs in a million years (Section 6.9).

A gram of rich soil may contain 10^9 microorganisms, rock presumably some orders of magnitude less. A single boulder landing on an external planet amounts to one unit in terms of spreading life; but if the boulder is eroded down to sub micron grains, 10^{15} of which contain microorganisms and which are dispersed in a biosphere expanding around the solar system, then there is clearly a potential for seeding of receptive environments on an interstellar scale. Thus for a given mass of ejecta, erosion enhances the capacity for spreading life by an enormous factor. This brings us to the second missing factor in the lithopanspermia analyses to date, close encounters with starforming regions.

Because of their great masses, GMCs have a disruptive effect on the Oort cloud, both throwing comets inwards towards the planetary system and ejecting them from the Oort cloud into the passing nebula. Since the solar system passes within 5 parsecs of starforming nebulae every ~50–100 My, a survivable distance for a protected microorganism, the question arises whether these starforming regions might be seeded by microbial life. This in turn raises the question of whether comets might act as 'messengers' transmitting information about life from system to system throughout the Galaxy.

To explore this further, we need to look more closely at the structure of the Oort cloud and how it is affected by the Galactic environment. Little is known about the internal constitution of the comet cloud, and much depends on the way in which comets were formed.

4.2 The Formation Sites of Comets

We have seen that comet chemistry points to a wide variety of formation origins, while the *Stardust* experiment showed that even the material inside a single comet, Wild 2, has come from diverse environments. The mineralogy of this comet shows a certain amount of organic material including, possibly, the amino acid glycine; but mainly there is an abundance of crystalline silicates comparable with those found in meteorites usually supposed to come from the asteroid belt. On the other hand, gas phase chemistry of several bright comets clearly points to an origin in an extremely low-temperature environment more like that of a molecular cloud than a protoplanetary nebula. This does suggest there may be a continuum of planetesimals from asteroids to comets, with a gradation of temperature and other properties in the planetesimal-forming regions.

If planetesimals were formed in a region of the protoplanetary nebula with chemical gradients, one might expect to use some measure such as the relative abundances of volatiles in comets to assess their sites of origin. However, no clear correlation has been discovered between the dynamics of comets and their volatile content. For example, the Halley-type comet Tuttle, with an orbital period of 13.6 years, has an abundance of volatiles which shows no systematic difference from that of Oort cloud comets (Bönhardt *et al.*, 2008). Significant mixing of Oort cloud, scattered disc and Edgeworth–Kuiper belt populations is in any case likely (Emel'yanenko *et al.*, 2007).

The history of mainstream ideas about the origin of comets is largely one of fine tuning various properties of the primordial solar nebula in an attempt to meet conflicting requirements. The problems involved are complex. We need to take account of the dynamics of grain growth as a result of mutual collisions and the differential settling of grains to form snowballs as the initial steps. Next we have to consider the aggregation of these snowballs through gravitational instability. All this happens in an environment where planets are growing, probably migrating as they do, and the Sun is generating an extremely strong solar wind during its T–Tauri phase.

There is a case for asserting that all comets are formed in the interstellar medium, since the protoplanetary disc collapsed out of it. The questions then to be asked are how much processing took place in the disc itself, how comets actually grew, and how the various comet populations got where they are. In particular, how did several hundred billion comets come to be placed in a spherical cloud stretching a third of the way to the nearest star?

4.2.1 Origin in the planetary region

Comet growth is widely considered to have occurred in the primordial solar nebula. This, it is often assumed, had a mass about 0.1 $M_\odot$, based on adjusting the observed planetary masses up to solar abundances. With an admixture of grains colliding randomly, and sticking when they did, it can be shown that kilometre-sized bodies could grow in the Uranus/Neptune region on a timescale of about a billion years. However if we assume a sticking probability say 5% then a comet of 10 km diameter would take about 4×10^{10} years to grow in the Neptune region, greater than the expansion age of the universe. If the origin of the outer solar system really was such a drawn-out affair, the Sun would have left its original star cluster before comets were formed.

These long timescales led Öpik, in a series of papers published in the 1970s (by which time he was in his 80s) to the view that the origin of comets could not lie any further out than the Jupiter–Saturn region, where the dust density was much higher and a 10 km body could grow in a few million years. He had to assume that during this period sunlight was greatly dimmed by a huge amount of intervening dust within the orbit of Jupiter, this dust persisting for some millions of years. More recent models of the same type that also take into account the perturbations of all the major planets, give an overall efficiency of about 10% for populating the Oort cloud — the other 90% of comets are either ejected into interstellar space or scattered inwards, ultimately to fall into the Sun.

Ernst Julius Öpik (1893–1985). Estonian astronomer who spent the last 33 years of his career at Armagh Observatory in Northern Ireland. He proposed that comets originate in a vast cloud orbiting far beyond the planetary system, a theory later adopted by the Dutch astronomer Jan Oort whose name is usually associated with the comet cloud. The existence of this cloud forms the basis of modern cometary dynamical studies (Courtesy: Wikimedia Commons).

Öpik then argued that Jupiter threw the comets so formed into the Oort cloud by gravitational slingshots. The efficiency of the process is low: even a slight excess of energy imparted would throw a comet into interstellar space. As a result, to match the known properties of the Oort cloud, Öpik had to assume that about 60 Earth masses of comets originally resided in the neighbourhood of Jupiter. This estimate is unquestionably much too low, since he greatly underestimated the population of the Oort cloud and also neglected post-formation losses.

The mass of the present day Oort cloud has been estimated at about 14 $M_{\oplus}$, which must be less — perhaps substantially less — than the original cloud. On this basis, and allowing for the perturbations of all the major planets, Bailey and Stagg (1988) estimated that the total initial mass of planetesimals must have been about 300 $M_{\oplus}$, an embarrassingly high number for standard solar nebula theory.

Jan Hendrik Oort (1900–1992). Dutch astronomer who was a pioneer of radio astronomy. His work led to the detection of the galactic halo, and made important contribution to studies of galactic structure. He followed Öpik in suggesting that comets came from a common origin of the solar system, defining the cloud of comets named after him (Courtesy: Wikimedia Commons).

Öpik's work brought out the problems involved in formation and emplacement of the comets. Modern research in this area makes full use of high-speed computers and orbital dynamics, and now takes account of the probability that Uranus and Neptune were formed much closer in than their present positions, since there was insufficient material to grow them *in situ*. In addition, since the Sun was probably born within a star cluster, the perturbing effect of neighbouring stars is also considered in models of Oort cloud formation.

However, it turns out that most of the problems faced by Öpik still persist. The perturbing effect of other stars in the assumed star cluster is to pull comets away from the planets and so soften the energy they inject. This enhances the efficiency with which the outer Oort cloud is populated, and generates a more tightly bound cloud (Fernández, 1997). Even so, taking the current outer Oort cloud to have a minimum mass of ~1 $M_\oplus$ (an order of magnitude less than the estimate of Bailey and Stagg) then it is found that the inner cloud must contain at least 10 $M_\oplus$. Then with 10% efficiency of emplacement from the Jupiter/Saturn region, the original planetesimal system in that locality must have had at least 100 $M_\oplus$ (Brasser *et al.*, 2008). This is likely to be a fairly severe lower limit, if only because close encounters with giant molecular clouds throughout the history of the solar system have surely led to a drastic depletion of the original cloud (Napier and Staniucha, 1982).

4.2.2 Origin in molecular clouds

Currently unsolved problems in comet cosmogony include the large mass required for the initial system of planetesimals, the growth of extremely fragile comets in what seems to have been a turbulent protoplanetary environment, and the rate of growth of planetesimals in whatever low-temperature environment they formed. The latter matters for panspermia because in the situation where it takes say 10^9 years for a comet to aggregate, many radioactive nuclides will have decayed and melting of the cometary interiors may not take place, except perhaps in exceptionally large bodies. A rapid collapse, on the other hand, would preserve ^{26}Al which, trapped in the interior of the comet, could melt the interiors yielding organic-rich liquid water for some millions of years.

In considering the growth and emplacement of comets, how far out from the Sun might we expect planetesimals to condense? Are we even confined to the protoplanetary disc? What about molecular clouds themselves as candidate sites?

Molecular clouds are observed to be highly structured on all scales. Direct observation of very small inhomogeneities is difficult, but CO studies carried out in a few accessible areas have revealed that they are composed of structures a few hundred AU across with $n_H \sim 10^3$ cm^{-3} or more, with a fractal structure (Heithausen, 2007). Theoretical considerations suggest that even finer structures should exist. Thus the passage of a strong shockwave, such as might be generated by a supernova explosion, into warm or cold neutral hydrogen generates a thin, dense layer of molecular hydrogen (Koyama and Inutsuka, 2000). Thermal instability within this layer creates unstable perturbations, whose shortest wavelengths are ~15 AU, generating H_2 cloudlets which collapse on timescales ~10,000 years. The dust content of such a structure has mass ~10^{21} g, corresponding to a body ~140 km across, about that of a Chiron-sized comet.

Inhomogeneities in the dust distribution may also be created by the action of radiation on grains, which tend to shield each other in an ambient radiation field, so creating a pseudo-gravity. Gerola and Schwartz (1976) pointed out that this tendency may be substantially enhanced by the photodesorption of molecules from the surfaces of icy grains, creating a rocket effect which would tend to drive grains together. Experimental (Westley *et al.*, 1995) and theoretical (Weingartner and Draine, 2001) investigations indicate that in a molecular cloud environment, asymmetric UV radiation is capable of accelerating a micron-sized grain exposed to radiation from OB stars across the dimensions of a diffuse cloud within its lifetime.

We can extend the condition for marginal instability, Eq. (3.1), to incorporate this pseudo-gravity:

$$2K_E + \Omega + \psi = 0 \tag{4.1}$$

where

$$\psi = \int_V \rho_g \boldsymbol{r} \cdot \boldsymbol{f} \, dV \tag{4.2}$$

is the radiative analogue of the gravitational potential, yielding an acceleration $\boldsymbol{f}$ on an individual dust grain, ρ_g the density of the grain. A perturbation of dust over some length will collapse when the differential radiation force within it yields an acceleration in excess of the thermal motion of the dust, ~1 cm/sec at 10 K.

Radiation pressure in a molecular cloud environment acts most effectively on grains of diameter 10^{-5} cm. The acceleration takes the form

$$f = \frac{\pi a^2 KUG}{m_g} \tag{4.3}$$

where πa^2 is the cross-section of a grain and m_g its mass. The factor K represents the momentum transfer from an impinging photon, which may be enhanced by photodesorption of molecules on the grain surface, U is the energy density of the ambient radiation field and G is a geometrical factor representing the anisotropy of the radiation field within the perturbation. This latter takes the form $G = \delta\tau$ for a homogeneous slab perturbation, or $G = \delta\tau/3$ for a spherical one, $\delta\tau$ the optical depth enhancement measured from the mid-plane or centre of the dust perturbation: it can be related to the enhancement δn_g in grain density through $\delta\tau = \pi a^2 \delta n_g r$.

Napier and Humphries (1986) find that, in a molecular cloud environment, the radiative analogue (Eq. (4.1)) of the gravitational Jeans equation (3.1) yields dust concentrations that are unstable to rapid collapse in the mass range 10^{15}–10^{21} g. This encompasses the entire range of masses appropriate to comets. Turbulent velocity shear, as for example measured in the dark clouds of the star-forming regions in Taurus, seems to be about two orders of magnitude short of that required to destroy such perturbations. It therefore seems that comets might grow out of the interstellar medium by aggregation of dust particles on a fairly short timescale.

Although, as we have seen, a great deal of theoretical effort has gone into examining how comets may be thrown out from the original planetary region into the Oort cloud, and this must indeed happen, the above considerations suggest that comets may form also — and perhaps even primarily — in regions much further out than the protoplanetary disc, perhaps even in the molecular cloud environment prior to star formation. One can then envisage a situation in which comets are both brought in with the collapsing gas and thrown out at a later stage of planet growth. Certainly this process would alleviate the difficulties which exist in forming cometary ices at temperatures more appropriate to a molecular cloud than the solar nebula, and in conditions far removed from thermodynamic equilibrium.

4.2.3 Exocomets

It seems unlikely that the formation of comets is peculiar to our solar system, if only because it would place humanity at a highly exceptional location. The fact that dusty discs are commonly found around main sequence stars, and that planetary systems seem very common, suggests that comets too, if formed along with planets, are ordinary constituents of exoplanetary systems. Evidence of comets around some other stars exists, although inevitably there is a layer of interpretation between observation and deduction.

When a star evolves off the main sequence, the large increase in its luminosity must lead to the vaporisation of any comets orbiting within several hundred astronomical units. This could be detected in the form of water vapour in the stellar spectrum of red giants. Abundant water vapour has been discovered around IRC+10216, an intensively studied star with a carbon-rich circumstellar envelope, which is in fact moving from the main sequence to its asymptotic giant branch. Equilibrium chemistry would tend to combine carbon in the disc with oxygen, leaving little to form water. The abundance of water, 10^4 times more than expected, seems to require the evaporation of about 10 $M_\oplus$ of icy planetesimals (Melnick *et al.*, 2001; Saavik Ford and Neufeld, 2001).

Another way to detect comets would be to look for absorption in the comae and tails of comets as they occlude the disc of the parent star. Variable, red shifted absorption lines have been detected in the spectrum of the red giant star β Pictoris and can only be explained in this way; they appear to be due to the sublimation of ices from large bodies as they approach the star (Vidal-Madjar *et al.*, 1994). A few other analogues of β Pictoris have been discovered.

Infrared satellite spectroscopy of 69 nearby solar-type stars has revealed evidence of debris discs in about 10–15% of cases, the emission from all but one coming from cool material located 10 AU or more beyond the stars. The one exception, the K0 star HD 69830, reveals spectral features such as crystalline silicates closely resembling those of comet Hale-Bopp occurring within a few AU of the star, and may be explained by the capture of a single giant comet into a short period orbit (Beichman *et al.*, 2005).

The actual frequency of cometary systems around other stars is probably high; this statement, however, owes more to theoretical preconception than observation. Given the inefficient emplacement of comets into the Oort cloud, it seems that interstellar comets must outnumber bound comets by a factor of 10^2 or 10^3, an excess which will hold *a fortiori* if comets have a molecular cloud provenance. No interstellar comets have yet been detected passing through the solar system, but two comets have been found with highly anomalous molecular abundances. Comet Yanaka is depleted in molecular carbon by a factor of 100 and in CN by a factor of 25 relative to typical comets, while in comet Machholz these molecules are depleted by factors of 8 and 70 respectively. It has been suggested that these comets are interstellar interlopers, coming from some interstellar cloud with different properties from those in which our sun was born (Schleicher, 2008). Capture from the interstellar medium, although very inefficient, could yield a small admixture of such bodies to the Oort cloud (Valtonen, 1983; Clube and Napier, 1984).

4.3 The Sun's Orbit in the Galaxy

The distance to the Galactic centre is 7.6±0.4 kpc, and the Sun orbits it at ~220 km/sec in a Galactic year of length ~220 million years. The peculiar motion of the Sun at the present epoch is ~16 km/sec relative to local stars. The one-dimensional velocity dispersion of molecular clouds is ~8 km/sec and their mean motion is drifting by ~4 km/sec relative to the local standard of rest (Stark and Brand, 1989). Boulders ejected from the solar system might therefore be expected to enter nearby (passing) molecular clouds and protoplanetary nebulae at speeds typically in the range 15–25 km/s. Such boulders would simply fly through the target cloud except in the rarest instances of planetary capture referred to above.

The transfer of viable biomaterial from boulders to the target system could take place in several ways, for example as discussed by Napier (2004) and Wallis and Wickramasinghe (2004). In the mechanism proposed by Napier grinding collisions occur in the zodiacal dust cloud in the plane of the solar system, and life-bearing dust is expelled by the action of radiation pressure (see the discussion in Chapter 6). This process ensures a fast and efficient transfer to a perturbing molecular cloud, the enhanced bombardment leading to boulder/dust expulsion being itself a consequence of the gravitational interaction between the molecular cloud and the Oort cloud of comets. Wallis and Wickramasinghe (2004) propose that larger life-bearing boulders released through impacts are more or less continuously added to the outer solar system from which they can be expelled. From here ~4 tonnes per year are expelled to reach passing molecular clouds and protoplanetary nebulae within which they become fragmented and stopped by frictional gas drag.

Without orbital perturbations, and with those long-period comets which enter the planetary system decaying, the flux of comets would decline to zero within a few million years. The loss cones — empty regions of orbital parameter space — which develop must be replenished. The disturbing forces which achieve this come from stars which penetrate the Oort cloud, from the Galactic tide, and from passing nebulae. The strongest perturbers acting on the Oort cloud are the vertical galactic tide (Byl, 1986) and passing molecular clouds (Napier and Staniucha, 1982). We now examine these.

4.3.1 The effect of the vertical Galactic tide

Because mass gradients are much steeper in the out-of-plane direction in the Galaxy, the strength of the vertical Galactic tide exceeds that of its horizontal component by an order of magnitude. To a good approximation, the tide exerts a restoring force per unit mass on an Oort cloud comet, relative to the Sun, given by

$$f(z) = -4\pi G\rho\Delta z \tag{4.4}$$

where ρ represents the local density of the Galactic disc at the height of the Sun, and Δz is the difference in the vertical height between Sun and comet. This may be written as

$$f(z) = -C^2\rho$$

where the Oort constant $C \sim 0.1$ kms^{-1} pc^{-1} with about 50% uncertainty.

Orbital motion in the presence of this restoring force may be found by classical procedures (e.g. Byl, 1986; Yabushita, 1989). Assuming a constant local density, energy is conserved and so the semimajor axis distribution remains unchanged; however the angular momentum cycles between components, and there exist both librational and precessional solutions for the perihelion distance of the comet and its Galactic latitude. The perihelion distance may, in the course of a precessional cycle, decline from something comparable with the semimajor axis down to a very small value, before increasing again, the circuit time for a long-period comet being typically ~0.2–0.6 Gy.

First-order perturbation theory yields an average change δq in perihelion distance q of a long-period comet over an orbit:

$$\delta q \sim 12.4(25000/a)^{-6.3 \pm 0.2} \tag{4.5}$$

where the semimajor axis a is measured in astronomical units. This is an order of magnitude greater than that caused by typical stellar perturbation:

$$\delta q \sim 0.66(25000/a)^{3.5}$$

The break-even point between tide and stars occurs at $a \sim 18500$ AU, where $\delta q \sim 1.9$ AU per orbit. Beyond this distance, the tide is dominant; within it, its influence rapidly fades relative to stellar perturbations. For a long-period comet whose perihelion has declined to q ~ 15 AU say, Eq. (4.5) tells us that on its next orbital pass the comet will have jumped over the 'Jupiter barrier' and entered the zone of visibility $q < 5$ AU. Having come and gone, it will never be seen again.

The upshot of this behaviour is that the loss cones in the outer Oort cloud are permanently filled, meaning that there will be a steady supply of long-period comets from the outer cloud. However at smaller distances a surge in the comet flux may be caused by some disturbance, say by a passing star. The loss cones so emptied then need to be repopulated. This is achieved by tides and stars acting in synergy (Rickman *et al.*, 2008).

4.3.2 Flux modulation due to the Sun's vertical motion

As the sun orbits the Galaxy, it bobs up and down in a carousel-like, nearly harmonic motion. For strictly harmonic motion, the half-period of the vertical oscillation is

$$P_{1/2} = \left(\pi / 4G\rho\right)^{1/2}$$

where ρ is the density of matter in the Galactic plane. With ρ in units of $M_\odot\,\mathrm{pc}^{-3}$, we have $P_{1/2} = 13.2\rho^{-1/2}$ My.

Corresponding variations in the flux of comets into the inner solar system are to be expected, reflecting the cyclic variations in density and hence tide (Napier, 1987; Matese *et al.*, 1995; Clube and Napier, 1996). The question arises whether any such periodicity could be detected in the record of impact craters on the Earth. Such a detection would put limits on the relative importance of asteroids and comets in the impact record. The possible relevance to panspermia lies in the organic richness of comets and in their much higher impact speed when they strike the Earth, resulting in a higher proportion of ejecta.

The Sun's vertical motion in the disc is easily calculated if the density distribution of ambient stellar and interstellar material $\rho(z)$ is known. The latter, taken from Hipparcos data (Holmberg and Flynn, 2004) is shown in Fig. 29 and yields a local in-plane density $\rho = 0.105\ M_\odot\,\mathrm{pc}^{-3}$. The average of many independent estimates is $0.15{\pm}0.01\ M_\odot\,\mathrm{pc}^{-3}$ (Stothers, 1998), while the highest estimates are around $0.2\ M_\odot\,\mathrm{pc}^{-3}$ (e.g. Bahcall,

1984). What matters in any comparison with impact cratering history is the central in-plane density time-averaged over the period concerned. For plausible Galactic models, this increases the mean in-plane density by 0–20% (Stothers, 1985). Taking the middle-of-the-road values, the half-period of the Sun's vertical oscillation over the last 250 million years was estimated by Stothers (1998) to be 37±4 My.

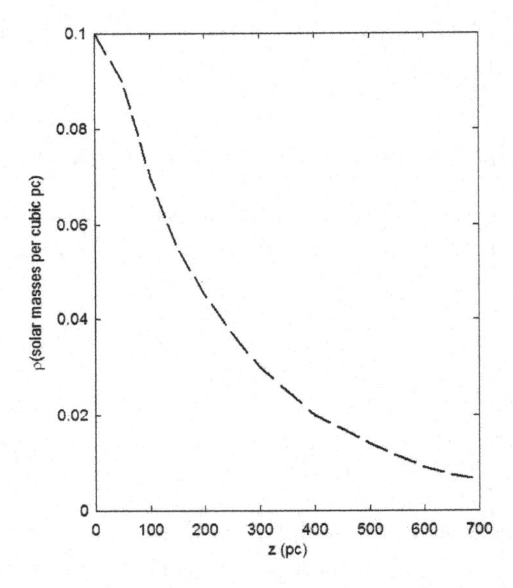

Fig. 4.1 The local density of galactic material (stellar and interstellar) as a function of height z.

The vertical motion of the Sun is then given by solving

$$\frac{d^2 z}{dt^2} = -4\pi G\rho(z) \tag{4.6}$$

and the period and amplitude of the solar orbit may then be found for a given value of the vertical velocity at $z = 0$. Figure 4.2 shows the numerical solution of this equation for the case $V_0 = 8$ km/sec and the $\rho(z)$ function defined by the curve of Fig. 4.1. Adopting this, the variation of flux of long-period comets into the solar system is derived as shown in Fig. 4.3. The tidal cycle has amplitude ~40% and the peaks are quite sharp.

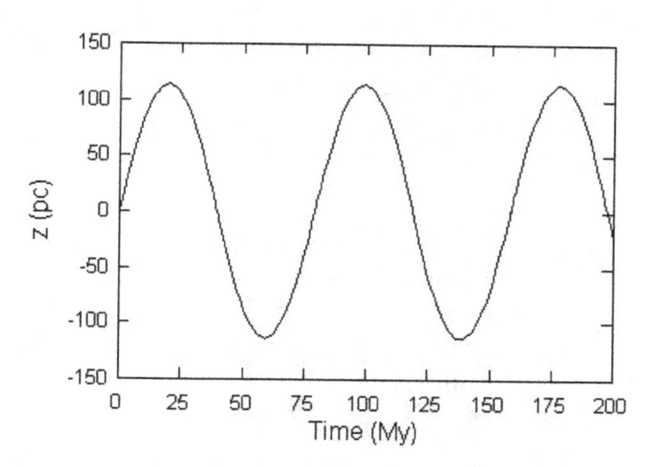

Fig. 4.2 Vertical motion of Sun with velocity of crossing the plane taken as 8 km/sec.

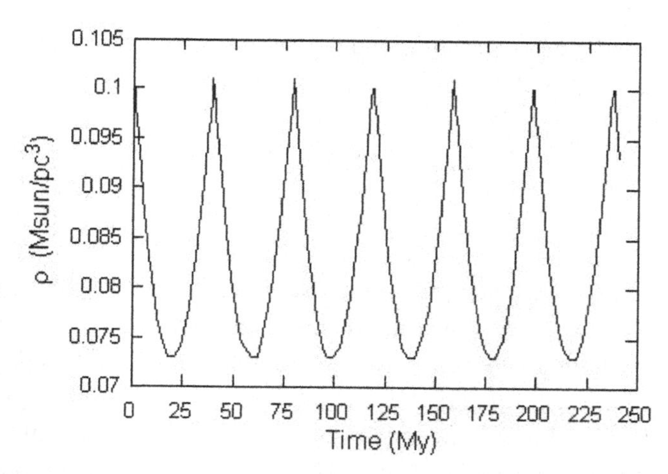

Fig. 4.3 Flux of long-period comets into the inner solar system using the Galactic disc density distribution of Fig. 4.2.

4.3.3 Perturbations by molecular clouds

In its orbit around the centre of the Galaxy, the solar system will from time to time encounter both giant molecular clouds and less massive dark cloud complexes. There are perhaps 4000 giant molecular clouds in the Galactic disc. Over the mass range $100<M_\odot<5\times10^5$, their differential mass distribution is given by

$$n(M) \propto M^{-\alpha}$$

with α = 1.6±0.2. The mass M of molecular clouds varies with radius as $M \propto R^2$ over at least eight decades of mass, whence small molecular clouds are denser. Thus the interval between encounters with nebulae of mass at least M is, in millions of years,

$$\Delta t \sim 800\left(\frac{M}{5\times10^5}\right)^{0.75}\left(\frac{d}{20}\right)^{-2}$$

(d in pc, M in $M_\odot$) when gravitational focusing is neglected.

Allowing for gravitational focusing, the effective interval is reduced by a factor

$$\sigma = 1 + (V_e / V)^2$$

where V represents the asymptotic encounter velocity and V_e is the escape velocity at the point of closest approach. This is a substantial factor in the case of a massive nebula: a GMC of mass 5×10^5 $M_\odot$ and radius 20 pc has a surface escape velocity V_e ~15 km/sec, and so for an asymptotic approach speed 15 km/sec the mean interval between grazing encounters is halved from 800 to 400 My. This implies that about ten such encounters may have taken place over the period during which life has existed on Earth. Figure 4.4 illustrates the mean intervals between encounters with nebulae of various masses and asymptotic approach speeds, for passages at 20 pc. It can be seen that encounters within 20 pc with molecular clouds of masses in the range 5000–10000 $M_\odot$ occur quite frequently on geological timescales (25–40 My).

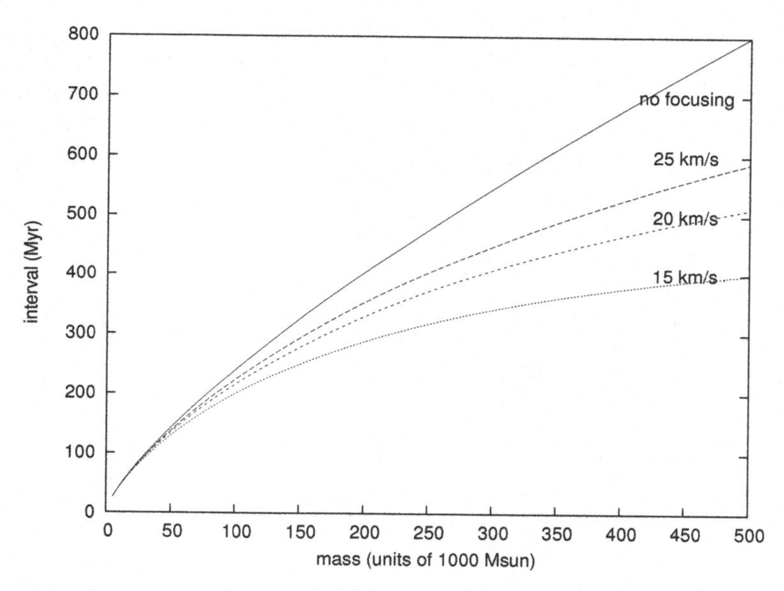

Fig. 4.4 Mean interval between encounters with nebulae for various asymptotic approach speeds, for impact parameter 20 pc, illustrating the effect of gravitational focusing.

4.3.4 The effect on Oort cloud comets

Discrete changes in the orbital elements of comets under the influence of the vertical Galactic tide can be computed using either numerical integrations or through analytic formulae which have been developed by Byl (1986), Fouchard (2004) and others. These formulae have been derived by averaging the perturbing force over an orbit and involve some approximations, but are orders of magnitude faster than direct integration.

Likewise the effect of a passing molecular cloud can be approximated using analytic formulae based on the impulse approximation. In situations where the comet orbits significantly while the perturbing nebula is still in the neighbourhood, the impulse approximation breaks down, but even then the trajectory of the nebula can be broken up into time steps and sequential impulses applied.

In the two-body problem, discrete changes of orbital elements can be calculated from formulae set out by Roy (1978) for a velocity impulse $\Delta \underset{\sim}{v} = (\Delta v_S, \Delta v_T, \Delta v_W)$, Δv_S, Δv_T, Δv_W being components of the impulse velocity relative to the Sun. Here v_S, v_T are the components of the comet's velocity in the orbital plane in the radial and transverse directions, and v_W is that normal to the orbit.

$$\left.\begin{aligned}
\Delta a &= \frac{2}{n\sqrt{1-e^2}}\left(e \sin f \Delta v_S + \frac{p}{r}\Delta v_T\right) \\
\Delta e &= \frac{\sqrt{1-e^2}}{na}\left[\Delta v_S \sin f + (\cos E + \cos f)\,\Delta v_T\right] \\
\Delta i &= \frac{r \cos u}{na^2\sqrt{1-e^2}}\Delta v_W \\
\Delta \Omega &= \frac{r \sin u}{na^2\sqrt{1-e^2}\,\sin i}\Delta v_W \\
\Delta \varpi &= \frac{\sqrt{1-e^2}}{nae}\left[-\Delta v_S \cos f + \left(1+\frac{r}{p}\right)\Delta v_T \sin f\right] + 2\sin^2\frac{i}{2}\Delta\Omega \\
\Delta \varepsilon &= \frac{e^2}{1+\sqrt{1-e^2}}\Delta\varpi + 2\sqrt{1-e^2}\,\sin^2\frac{i}{2}\Delta\Omega - \frac{2r}{na^2}\Delta v_S
\end{aligned}\right\} \quad (4.7)$$

The quantity f is the true anomaly, $p = a(1 - e^2)$, n is the mean motion such that $n^2a^3 = G(M+m)$ and $u = f + \varpi - \Omega = f + \omega$. r is the heliocentric distance given by

$$r = a(1 - e\cos E) \quad (4.8)$$

where E is the eccentric angle. True anomaly f and eccentric angle E are calculable at a prescribed time from the standard formulae of the two-body problem (Roy, 1978). The instantaneous acceleration of the comet relative to the Sun is along the radius vector: thus $\Delta v_T = 0 = \Delta v_W$. At every time step, with Δv_S calculated as a gravitational radial impulse, Eqs. (4.7) and (4.8) may be used to change the orbital elements for every comet in a given set (Wickramasinghe, J.T., 2007).

Such numerical tricks allow one to follow the orbital evolution of 10^5–10^6 Oort cloud comets under the combined influence of Galactic tide and passing nebulae. Figure 4.5 shows the flux of comets entering the planetary system, taken as a sphere of radius 40 AU, consequent on a grazing encounter with a GMC. The flux in this case was computed by direct numerical integration of 90,000 orbits whose initial semimajor axes were distributed as $n(a) \propto a^{-\gamma}$, with $\gamma = -2$. In this trial the Galactic tide was held constant and a comet, having entered the planetary system, was taken to have a survival probability 50%. This assumes equal chances of being gravitationally perturbed by Jupiter and Saturn into a hyperbolic orbit so ejecting it from the solar system, or being flung inwards and destroyed (Fernández, 2005).

We see a strong bombardment episode, declining with a half width ~3 My. A significant feature of such encounters is that the flux of comets into the planetary system increases for several million years while the GMC is still approaching. Thus an episode of bombardment is underway before the nebula reaches its closest point to the solar system. More centrally condensed Oort clouds yield stronger surges; for example with $\gamma = -3$, the comet flux peaks at 30 times the general background. The night sky for some millions of years during such an encounter would be filled with comets, while half the sky would be empty of stars.

Encounters with nebulae of smaller mass are more frequent, but are of less consequence from the perspective of disturbing the Oort cloud. Figure 4.6 illustrates the effect of encounters with a 50,000 solar mass nebula, at 20 pc and 5 pc respectively. The more distant encounter temporarily enhances the comet flux by a factor of three or so. The encounter at impact parameter 5 pc has a much more drastic effect, but such close passages are relatively infrequent, occurring at only ~2 Gy intervals.

In these simulations a slow secular decline of comets with time can be seen over the 40 My integration period. This arises because perturbations by passing stars, which should strictly have been taken into account as a background effect, were neglected. Moreover the refilling of emptied loss cones, by Galactic tides alone, is slow and inefficient, as pointed out by Rickman *et al.* (2008).

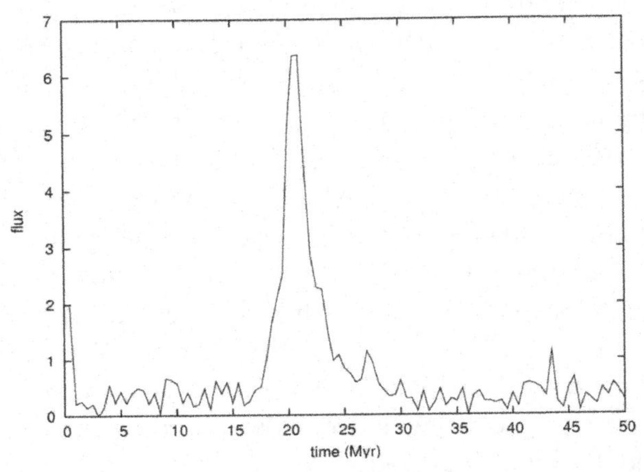

Fig. 4.5 A 90,000 particle simulation of a grazing encounter with a GMC of mass 5×10^5 $M_\odot$, radius 20 pc. Outwith the encounter, the flux of comets into the planetary system is determined by the Galactic tide. Note that this flux increases significantly before the time of closest approach at 20 Myr.

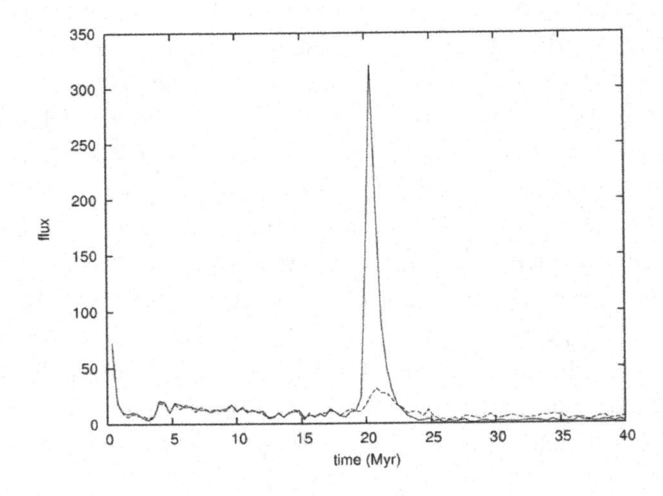

Fig. 4.6 Encounters with a 50,000 $M_\odot$, at 20 pc (lower curve) and 5 pc (upper curve) respectively.

4.4 The Impact Cratering Record

It is popularly supposed that the largest terrestrial impact craters on Earth were caused by the collision of asteroids. This is unlikely to be the case in any great measure. An asteroid in the main belt may be injected into a resonant orbit and, through pumping by planetary perturbations, find itself eventually in an orbit which makes it a collision hazard. The procedure often takes tens of millions of years, depending on the resonance involved. This is probably an effective source of sub-kilometre bodies, since these are relatively easy to dislodge by impacts into a resonant zone. However the more massive the asteroid, the more difficult it is to dislodge by collision into such regions and the process becomes inadequate above 1 or 2 km. The largest impact craters are much more likely to have been caused by the impacts of comets. With the Oort cloud as a major supplier, either directly or through intermediate comet populations, then we can ask whether the character of the impact cratering record really does reflect the expectations of an Oort cloud subject to Galactic perturbations. If so, then the system of long-period comets becomes a crucial link between the Galaxy and the Earth, their impacts being the way in which each talks to the other. We therefore have to ask whether the impact cratering record shows evidence of Galactic modulations, namely sharp bombardment episodes, possibly with some sign of periodicity.

The 200 or so known terrestrial impact structures are heavily weighted towards the Canadian and Baltic shields, and the desert areas of the Earth. In the whole of India, China, Tibet, Japan and Indonesia, there is only one modest sized crater, Lonar, in central India, 50,000 years old and already heavily eroded. It will be gone in the blink of a geological eye. Given that two thirds of the Earth's surface is water, it is clear that the impact cratering record must be seriously incomplete. This can be seen too in the age distribution of craters. Taking the twelve largest craters known, over 40 km in diameter, well-dated ($\sigma < 2.6$ My), with ages <250 million years, we find the distribution shown in Fig. 4.8. Sedimentation and erosion are serving to remove even the largest terrestrial impact craters. Probable bombardment episodes are listed in Table 4.1.

Table 4.1 Probable bombardment episodes of the past 250 My. For the most part impact craters listed have radiometric ages dated to better than ~5 My (see text for precision of Rochechouart age). Data from Earth Impact Database maintained by Planetary and Space Science Centre, University of New Brunswick, Canada.

Episode	Crater	Age (My)	Diameter (km)
1	Zhamanshin	0.96 ± 0.1	14
	Bosumptwi	1.1 ± 0.0	10
	Kara-Kul	2.5 ± 2.5	52
	El'gygytyn	3.5 ± 0.5	18
	Bigach	5.0 ± 3.0	8
	Karla	5.0 ± 1.0	10
2	Chesapeake	35.5 ± 0.3	90
	Popigai	35.7 ± 0.2	100
	Mistastin	36 ± 4	28
	Wanapetei	37.2 ± 1.2	8
3	Ragozinka	46 ± 3	9
	Gusev	49.0 ± 0.2	3
	Kamensk	49.0 ± 0.2	25
	Montagnais	50.5 ± 0.76	45
4	Chicxulub	64.98 ± 0.05	170
	Boltysh	65.17 ± 0.64	24
5	Ust-Kara, Kara	70.3 ± 2.2	25, 65
	Lappajarvi	73.3 ± 5.3	23
	Manson	73.8 ± 0.3	35
6	Mjolnir	142.0 ± 2.6	40
	Gosses Bluff	142.5 ± 0.8	22
	Morokweng	145.0 ± 0.8	70
7	Zapadnaya	165 ± 5	3
	Puchezh-Katunki	167 ± 3	80
8	Manicougan	214 ± 1	100
	Rochechouart	214 ± 8	23

Photo 4.1 As of 2009, 176 craters have been identified on the Earth, locations indicated by dots on the above map. Below is a montage of craters varying in diameter from between 0.3 km to ~100 km (Courtesy: NASA/JPL, NASA/LPI and the US Geological Survey).

One might question whether any useful information could be extracted from such a desperately small and incomplete sample. Nevertheless, patterns do emerge. Figure 4.7 represents a compilation of 40 impact craters over 3 km across and with radiometric ages measured to $\sigma < 10$ My (most of them <5 My). The impression that craters are bunched in age, interspersed with quiescent periods, is verified by formal statistical analysis (Napier, 2006). The impact epoch occurring at the present time appears to be real and not a discovery artifact.

In assessing these putative bombardment episodes, several factors need to be taken into account. It might be assumed that the first episode creates the illusion of bunching simply because discovery of impact craters is more complete if the craters are very young. However this cluster of impact craters is quite distinct: there is a long gap between it and the next one, with only one modest impact structure (Ries, in Germany, at ~15 My) in between. Episode two straddles a series of severe extinctions and climatic downturns at the Eocene–Oligocene boundary. Episode three has perhaps the weakest supporting evidence; two of the impact craters listed (Kamensk and Gusev, in Russia) are very close together — Gusev is likely a secondary crater — and so can only be counted as one object in any statistical assessment. Episode four includes the Chicxulub crater in the Mexican Gulf associated with the dinosaur extinctions. The occurrence of a substantial crater (Boltysh) in the Ukraine, with identical age to a high degree of precision, clearly indicates that a multiple bombardment event took place at the Cretaceous–Tertiary boundary.

As we go into the more remote past, discovery becomes seriously incomplete and it is all the more remarkable that three bombardment episodes (6, 7 and 8) can be recognised. The age uncertainty given for the Rochechouart crater is given as ±8 My, but debris probably from this impact is to be found in southwest England and has been dated to ±2.5 My, putting it extremely close in time to the Canadian Manicouagan crater.

The bombardment episodes are strong and all of the larger impact craters ($D > 40$ km) belong to them. It seems that the bombardment episodes are short in duration but intense, involving several bolides 2–5 km or more across as well as, presumably, dozens of lesser missiles. The breakup of main belt asteroids and the feeding of fragments into Earth-crossing orbits generally cannot yield the sharpness and amplitude observed (Zappalá *et al.*, 1998; Napier and Asher, 2009).

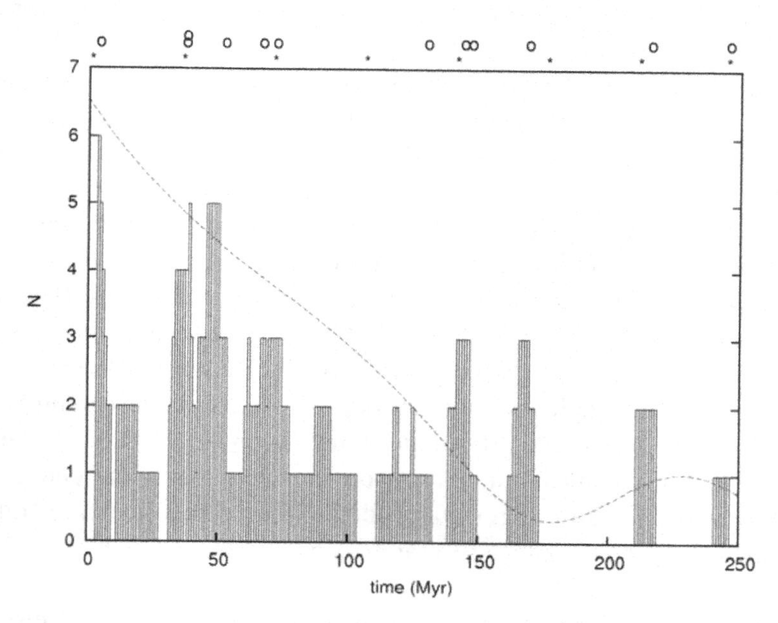

Fig. 4.7 The age distribution of 40 impact craters 3 km or more in diameter, with ages less than 250 Myr known to precision better than 10 Myr (data smoothed in this plot by a window of width 8 Myr). Impacts occur in discrete episodes of bombardment. The circles represent the formation date for 12 craters over 40 km across with ages measured to precision 2.6 Myr or better. The asterisks mark out a best-fitting periodicity of ~35 Myr for those 12, and the smooth curve is a cubic spline fit for all 40 craters.

4.4.1 Impact melts in large craters

A possible counterargument to the hypothesis that these are cometary bombardment episodes involves the impact melts in two of the largest craters, the Russian Popigai crater at ~36 My and the South African Morokweng crater at 145 My. Both of these have impact melt material within them with compositions matching those of chondritic meteorites. This has led to the proposal that the corresponding bolides were asteroidal rather than cometary.

Almost nothing is known about the range of cometary compositions and how it compares to the range of asteroidal materials. However partially melted chondrules — essentially meteoritic fragments — have been found inside the Morokweng crater, going down to the base of the melt sheet at 900 m depth.

Chondrules are spherical particles, in the submillimetre to mm size, which were once molten silicate drops, having been heated within a few minutes before cooling rapidly. Their origin is uncertain, but recent numerical modelling has indicated that low-velocity impacts on to porous bodies may produce a substantial volume of chondritic material. They are commonly found in meteorites and are amongst the most primitive material in the solar system. It is possible that comets with a high mineral content, assembled in the nascent planetary system, will contain chondrules either gathered up during accretion or formed *in situ* within them. Comets have been proposed as the source of certain types of chondritic meteorite (Campins and Swindle, 1998). Detailed petrologic and chemical studies support the view that some chondritic meteorites are actually fragments of long-extinct comet nuclei (Lodders and Osborne, 1999). Likewise grains recovered from comet Wild 2 by the *Stardust* mission had mineral assemblages (olivine, pyroxene, sulphides etc.) typically associated with some chondrite types (Weisberg and Connolly, 2008).

The old sharp division of interplanetary bodies into rocks (asteroids) and dirty snowballs (comets) seems no longer to be valid and it is now generally recognised that there is a gradation of internal properties as well as many transitional objects. It is likely, too, that an admixture of rocky asteroids was ejected into the Oort cloud along with icy bodies. At present, therefore, it is not clear just what is being constrained by the evidence from the impact melts in Popigai and Morokweng.

An odd feature of the Morokweng crater is that it contains an extremely high amount of chondritic material — almost as if the projectile had been lowered into the crater! A low asymptotic approach speed to the Earth is implied, perhaps with a grazing entry; at any rate, the impactor must have had an unusual orbit.

4.4.2 Galactic periodicity

The question arises whether the structure of the bombardment episodes is consistent, at a quantitative level, with our model involving Galactic perturbations of the Oort cloud, yielding impacts either directly through long-period comets or indirectly through subsidiary populations such as the Halley-type system. In particular, we may ask whether there is evidence of Galactic periodicity in the record, arising from a periodic variation in the forces acting on the Oort cloud as the Sun orbits the Galaxy.

Given that the cratering dataset will have a large random component due to asteroids straying from the main belt, and that the flow of comets from the Oort cloud will itself contain sporadic elements, it may be surprising that any periodicity at all would be detectable. Nevertheless such claims have been made frequently over the past 20 years, and as the dataset has grown a pattern has emerged. Napier (2006) and Stothers (2006) have pointed out that the claims of periodicity tend to fall into two groups, one centred around 27 My and the other around 35 My, with the 35 My periodicity being concentrated in the largest craters, say over 35 km across.

The Sun is 9±4 pc north of the Galactic plane which, with the current vertical velocity 7 km/sec, means that we passed through the plane about 1 My ago. Allowing for the infall time of long-period comets, it seems that we should be in the peak of a bombardment episode now, consistent with the current impact surge (Fig. 4.7 and Table 4.1).

The motion of the Sun around the Galaxy has been modelled taking account of both the vertical tide and sporadic encounters with molecular clouds, assumed to lie in the plane with half height 50–60 parsecs. The fluctuating infall of long-period comets was computed, and synthetic datasets containing 12 craters extracted randomly from this variable flux. Standard statistical procedures were then applied to see whether the inbuilt Galactic periodicity could be retrieved. Account was taken of the increasing difficulty of discovering impact craters with time (Fig. 4.8).

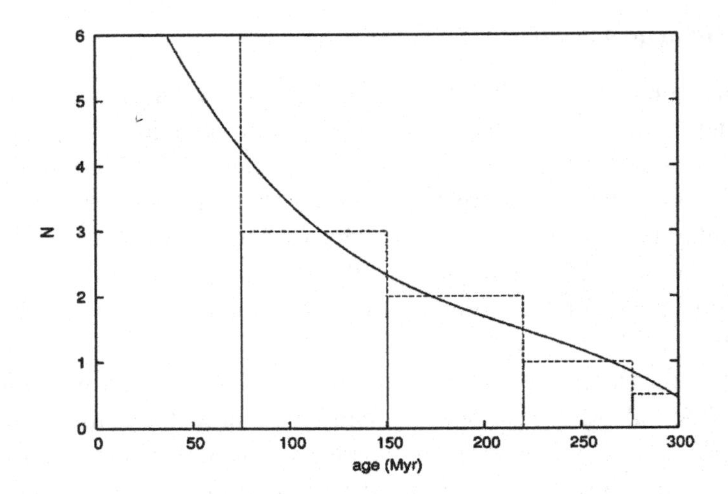

Fig. 4.8 The age distribution of 12 large, well dated impact craters of the last 250 Myr. Diameters > 40 km and ages accurate to better than 10 Myr.

Figure 4.9 shows the outcome of bootstrap analysis (5000 trials) applied to one such synthetic dataset. These simulations demonstrate that the inbuilt periodicity of ~35 My is quite well retrieved, although some solutions (depending on the vagaries of data selection) yield harmonics. Quite often in these simulations, a 27 million year harmonic appears. This periodicity has often been claimed in the literature, and its retrieval in models with an inbuilt 35 My cycle may provide a clue to why this is so. In the case of a weak dataset, harmonics sometimes predominate, depending on chance.

Figure 4.10 shows the outcome of applying this periodicity-hunting procedure to the 12 largest, well-dated impact craters. The smaller craters may well be dominated by asteroids from the main belt thrown into near-Earth orbits, and the inclusion of this large random component will tend to weaken the underlying signal when all craters, large and small, are lumped together. Remarkably, however, when we concentrate on the small number of large craters, a stable periodicity of ~35 My emerges. It is also very satisfactory that the near-zero phase of this periodicity is consistent with our expectations, given that the Sun has just passed through the plane of the Galaxy.

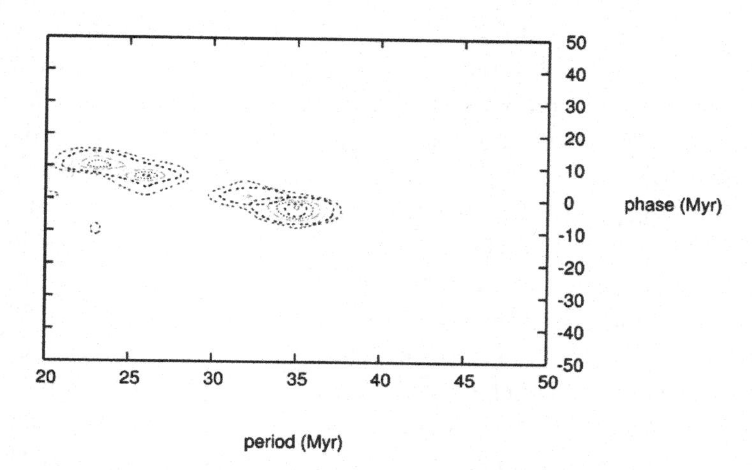

Fig. 4.9 Retrieving the periodicity from synthetic data. The motion of the Sun around the Galaxy has been simulated adopting a central plane density 0.16 $M_{\odot}$ pc^{-3}. Synthetic impact craters are extracted from the dynamical model and analysed for periodicity (de-trending and applying power spectrum analysis to bootstrapped data). The inbuilt periodicity in the model is well retrieved (P ~ 35 Myr, φ ~ -2 Myr). Phase φ is defined as the time elapsed since the most recent episode.

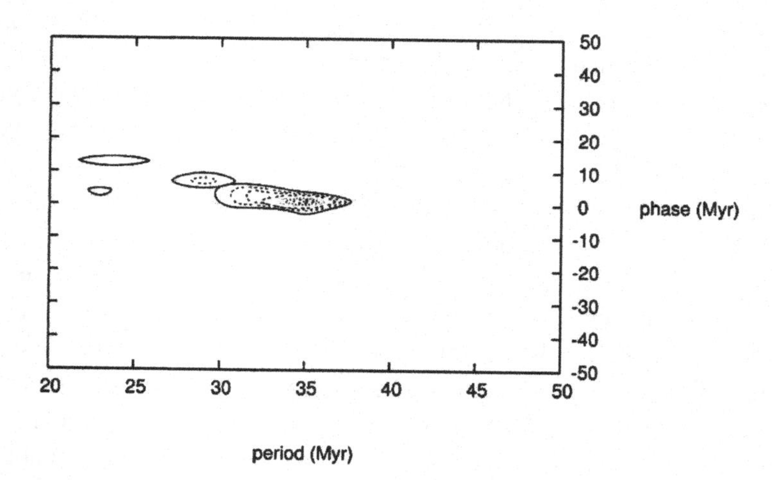

Fig. 4.10 The real data. A periodicity ~35 Myr, with phase close to zero corresponding to our current passage through the Galactic plane, is seen in the 12 impact craters >40 km across. This matches the Sun's vertical motion provided that the central plane density over the past 250 million years is ~0.16 $M_{\odot}$ pc^{-3}.

In conclusion, we find that both the presence of strong impact episodes and the existence of periodicity in the impact cratering record are difficult to reconcile with an asteroidal source. Rather, they point to cometary bombardment episodes as the major inputs to the terrestrial cratering record. The threshold crater diameter at which the periodicity emerges corresponds to impactors of 1.5–2 km in diameter. (This incidentally, is the threshold at which an impact would probably lead to global catastrophe, with a significant part of humanity destroyed.) Above this, comets seem to predominate; below that, the bulk of the impactors are probably asteroidal. The disintegration of large comets regenerates the zodiacal cloud and ensures an infall of dust to the Earth in the form of meteors, both sporadic and in the form of annual showers. These small particles enter the Earth's atmosphere with brief flash heating, which any microorganisms they contain may survive.

Conversely, comet impacts have the effect of transferring rocks and boulders bearing microbial life from Earth to interplanetary space, where they become collisionally ground and fragmented into dust that can be expelled by radiation pressure into nearby interstellar space. The passing molecular cloud, which caused the bombardment episode in the first place, would have its share of nascent planetary systems within which microorganism-bearing dust particles would become trapped. In Chapter 6 we discuss the question of survival of these tiny space travellers. We show that a fraction adequate for our purposes will always survive this short interstellar hop.

Thus panspermia outwards from Earth appears to follow logically from the known distribution of star-forming clouds in the Galaxy, combined with well-attested celestial mechanics. Since Earth cannot be considered the 'centre' of any such process, and because Earth-like planets are thought to be commonplace in the Galaxy, exchanges similar to those discussed in this chapter must also occur on a Galactic scale. On this view the microbial and evolutionary life legacy of our Galaxy is thoroughly mixed, with DNA-based life occupying every habitable niche.

Most incoming Oort cloud comets will probably not strike the Earth directly. Rather, they will transfer into one of the intermediary populations such as the Chiron, Jupiter-family or Halley-type systems.

We shall now examine a curious problem which arises in connection with this transfer.

Chapter 5

Dark Comets: A Link to Panspermia

The long-period comets entering the inner solar system arrived from an immense reservoir, the Oort cloud, which comprises ~5×10^{11} comets (down to 1 kilometre in diameter) orbiting the Sun. The arriving comets have aphelia in the range 3,000–100,000 AU, with a peak at about 33,000 AU when plotted in intervals of equal energy, $1/a$. This probably reflects the region of greatest efficiency of perturbation of the comets rather than a true peak in number density.

As we saw in Chapter 4, the long-period comets are driven inwards primarily by Galactic tides acting in synergy with stellar perturbations. Sporadic encounters with molecular clouds and stars during the solar system's motion through the Galaxy also give rise to discrete episodes of comet injection into the inner regions of the solar system on timescales of ~30–100 My.

It is possible that a dense inner cloud, impervious to tides, does exist. Over geological timescales, rare passages of stars through such an inner cloud would yield strong comet showers, showing temporal clustering in the impact cratering record. Study of the record of impact cratering can therefore constrain the number density of any such inner cloud. At its extreme, disturbance of a dense inner cloud would yield a night sky filled with comets. At the present day, however, this inner Oort cloud is inactive — if indeed it exists.

Some proportion of the arriving long-period comets will be perturbed by the giant planets into intermediate populations. An interesting paradox arises when we consider this transfer, in particular the replenishment of the system of Halley-type comets. To describe this paradox, we first consider the rate of arrival of long-period comets into the planetary system.

5.1 A Mass Balance Problem

The absolute magnitude H of an active comet (nucleus and coma) is defined as its visual magnitude if placed at 1 AU from the Sun and 1 AU from Earth. Emel'yanenko and Bailey (1998) estimated, from observations, that the present day flux of new, near-parabolic comets brighter than $H = 7$ is ~0.2 comets per annum per AU. Thus a comet of at least this brightness will pass within 1 AU of the Sun every five years and have a high probability of being discovered.

Going down to $H = 10.5$ or fainter, Fernández (2005) estimates that about 0.8 comets pass in a unit perihelion interval per annum. These numbers are derived from comet observations covering a long time span, but seem quite robust and are close to those derived using only *Spaceguard* observations (Francis, 2005), where observational selection effects are better understood. Roughly speaking, an active comet of absolute magnitude 7 has diameter ~10 km, one of absolute magnitude 10.5 has diameter ~1 km.

Taking account of planetary, stellar and Galactic perturbations, the flux of ~10 km-sized near-parabolic comets from the Oort cloud into perihelion distances $q < 4$ AU turns out to be approximately one per annum. It turns out that the Oort cloud plays a fundamental role in replenishing the various cometary reservoirs. Numerical studies support the view that the Oort cloud is the source of the majority of comets passing through the inner planetary system, even when arriving through intermediary populations such as Jupiter-family or Halley-types (Emel'yanenko *et al.*, 2007). Thus once a long-period comet arrives in the region of the giant planets, its dynamical evolution becomes subject to their gravitational influence, and there is a finite probability that it will be perturbed into Halley-type and other orbits. It is possible to calculate the proportion of comets that are thrown into 'Halley-type' orbits (with periods $20<P<200$ years, and perihelia $q < 4$ AU, say). Most of them originate from initial orbits with small perihelion distance, say $q < 4$ AU.

Emel'yanenko and Bailey (1998) found that in the perihelion range $0<q<4$ AU, the probability of being captured from the Oort cloud into a Halley-type orbit was ~1%, rising to ~2% or more if non-gravitational forces were allowed for.

Adopting the 1% estimate and assuming a life-time of a Halley-type comet as being τ yr, there should be a steady-state population of

$$N = 0.01\,F\tau$$

comets brighter than $H = 7$ in Halley-type orbits, where F is the flux of long-period comets into the range $0<q<4$ AU. Numerical integrations reveal that the dynamical lifetime of a comet in a Halley-type orbit, although highly variable, has a mean in the range 300,000–500,000 years. Its fate is to strike a planet, fall into the Sun or be ejected into interstellar space (the physical lifetime may be less than the dynamical lifetime; here we are considering the latter). With $\tau = 300{,}000$ yr and $F = 1$/yr, there should be about 3000 comets in Halley-type orbits at any given time. And this is the paradox: this theoretical number is over 100 times greater than the observed number. If we go down to say kilometre-sized comets, the mismatch between theory and observations becomes a factor of hundreds. This severe discrepancy has been confirmed in several independent studies (Levison *et al.*, 2002; Biryukov, 2007).

A number of ways in which this paradox might be resolved have been discussed in the literature. The comets may disintegrate to dust; or they may become dormant, developing dark or superdark mantles, becoming in essence stealth comets observable only in the thermal (satellite) infrared. Whatever the resolution, the imbalance implies that an incoming Halley-type comet has a short lifetime as a visible object.

5.1.1 Disintegration to dust

Comets lose gas and dust and their physical lifetimes may well be shorter than their dynamical ones. Levison *et al.* (2002) proposed that at least 99% of Halley-type comets disintegrate completely during their first few perihelion passages, stating that a comet evolving into an orbit with $q < 1$ AU has a 96% chance of disrupting before its next perihelion passage.

In fact the comprehensive disintegration of a 10 km comet into dust within a couple of revolutions does not seem likely. Halley's Comet itself has been reliably observed for almost thirty revolutions, and nearly all the major meteor streams, which are recorded in Chinese and Korean annals going back for one or two millennia, have at least one large body embedded within them. Disintegration by a hierarchy of splitting into small fragments seems to be as common a fate for comets as their steady disintegration into dust. Unfortunately the lifetime of such fragments is completely unknown, and one cannot say whether or not they would end up as dust within the dynamical lifetime of the Halley system.

Consider firstly the 3000 bright comets (with $H < 7$) which are supposed to be resident in Halley-type orbits but to have disintegrated to dust. The differential mass distribution of long-period comets may be fitted by a power law

$$n(m)\,dm \propto m^{-\alpha}\,dm \tag{5.1}$$

where $\alpha \sim 1.47$ (Weissman and Lowry, 2001). This does not differ significantly from the best-fit $\alpha \sim 1.38$ obtained from the few known objects in Halley-type orbits (Levison *et al.*, 2002). In the calculation that follows we use $\alpha = 1.4$. A least-squares fit to the masses, m, and absolute magnitudes, H, of 27 long-period comets is given by

$$\log m = 21.13 - 0.48\,H \tag{5.2}$$

assuming a mean albedo $p = 0.05$ and density $\rho \sim 0.5$ g cm^{-3} (Bailey, 1990). This yields $m_1 \approx 5.9 \times 10^{17}$ g for a comet with magnitude $H = 7$. If we take the brightest comets in our set to have $H = H'$, say, then the upper limit of mass is given by

$$\log m_1 = 21.13 - 0.48H' \tag{5.3}$$

and the average mass is

$$\overline{m} = \int_{m_1}^{m_2} m m^{-\alpha}\, dm \Big/ \int_{m_1}^{m_2} m^{-\alpha}\, dm$$

$$= \frac{m_2^{2-\alpha} - m_1^{2-\alpha}}{m_2^{1-\alpha} - m_1^{1-\alpha}} \cdot \frac{1-\alpha}{2-\alpha} \tag{5.4}$$

using Eq. (5.1). With $\alpha = 1.4$ and the lower limiting magnitude set at $H' = -1.5$, say, corresponding to a comet with diameter ~100 km, we obtain $\overline{m} = 1.1\times10^{20}$ g.

We now consider a sample of 3000 comets randomly extracted from this set. With an average mass ~10^{20} g, the total mass for the disintegrated Halley-type system is ~$3000\times10^{20} = 3\times10^{23}$ g.

In a typical case, taking period $P = 80$ yr, semimajor axis $a = 18$ AU and eccentricity $e = 0.96$, we find that the apsidal angles for which $r \leq 4$ AU lie in the range ±131°. In this case the time that the cometary material spends within 4 AU of the Sun is ~1.6 yr. The mass of material within this sphere is of order $3\times10^{23}\times1.6/80$ ~6×10^{21} g, which is 30–300 times the estimated total mass of the zodiacal cloud within 3.5 AU (Hughes, 1996). Such a dust sphere would create a strong visible glow in the night sky, but this is not observed.

En route to joining the dust sphere, debris from the postulated disintegrations would form streams. Some of these would intersect the Earth to produce annual meteor showers. Levison *et al.* (2002) predicts that there would be 46,000 dormant and active Halley-type comets with $q < 1.3$ AU and absolute magnitudes (nucleus only) $H < 18$ corresponding to diameters >2.4 km. If we take 30,000 Halley-type objects with $q < 1.0$ AU and mean periods 60 yr, then we expect 30,000/60 = 500 such bodies to pass within the Earth's orbit each year, and ~100 nodes of erstwhile dormant comets to come within ±0.05 AU of the Earth's orbit. The characteristic width of a strong meteor stream is ~0.05 AU, and so the number of recognisable streams from disintegrated comets is of order

$$N \sim 100 L_m / L_c \tag{5.5}$$

where $L_c \sim 100$ ky is the mean dynamical lifetime of the dormant comets and L_m is that of their associated meteor streams.

To estimate the mean lifetime of the meteor streams, we carried out simulations in which comets were randomly extracted from the 'dormant comet' parameter space computed by Levison *et al.* (2000). Each comet was broken into 27 pieces at perihelion, representing the meteoroids. These fragments were given random speeds up to δV in random directions. Laboratory experiments on simulated comet nuclei indicate that the ejection speeds of dust particles in the range 1–100 microns are a few meters per second when irradiated at a heliocentric distance $\lesssim 1.5$ AU (Ibadinov, 1989).

By adapting orbital integration programs developed by Chambers and Migliorini (1997) it was possible to follow the fragments for 100,000 years or until they had fallen into the Sun, collided with a planet or been hyperbolically ejected.

We measured the coherence of the meteor streams using a standard similarity function defined by Southworth and Hawkins (1963). For two orbits A, B with Keplerian orbital parameters defined by the 5-vectors $O_k = \{q, e, \omega, \Omega, i\}_k$, $k = A, B$, the orbital similarity function D is defined as

$$D^2 = [e_B - e_A]^2 + [q_B - q_A]^2 + \left[2 \cdot \sin\frac{I_{BA}}{2}\right]^2 + \left[\frac{e_B + e_A}{2}\right]^2 \left[2 \cdot \sin\frac{\pi_{BA}}{2}\right]^2 \qquad (5.6)$$

Here I_{BA} is the angle made by the orbit's planes given by the formula

$$\left[2 \cdot \sin\frac{I_{BA}}{2}\right]^2 = \left[2 \cdot \sin\frac{i_B - i_A}{2}\right]^2 + \sin i_A \sin i_B \left[2 \cdot \sin\frac{\Omega_B - \Omega_A}{2}\right]^2 \qquad (5.7)$$

and π_{BA} is the difference between the longitudes of perihelion measured from the points of intersection of the orbits,

$$\pi_{BA} = \omega_B - \omega_A$$
$$+ 2 \cdot arc\sin\left[\cos\frac{i_B + i_A}{2} \cdot \sin\frac{\Omega_B - \Omega_A}{2} \cdot \sec\frac{I_{BA}}{2}\right] \quad (5.8)$$

where the sign of the arcsin is changed if $|\Omega_B - \Omega_A| > 180^\circ$. The quantity D for our set of 27 meteors is calculated by averaging $<D_{AB}>^2$ given by Eq. (5.6) for all distinct pairs, and computing

$$D = \sqrt{<D_{AB}>^2} \quad (5.9)$$

Figure 5.1 shows a typical case of a comet with q = 1 AU, e = 0.96, $i = 152^\circ$. The bottom three panels show the evolution of e, q and i for the 27 cometary fragments, and the upper panel shows the evolution of a coherence parameter defined by Eqs. (5.6)–(5.7).

The known strong meteor streams have $D \lesssim 0.2$. We found that a significant proportion of the meteor streams in our calculations remained coherent throughout their evolution. Adopting $D_{crit} = 0.2$ as the cut-off for recognising a strong meteor stream, Fig. 5.2 shows the lifetimes of meteor streams for initial velocity dispersions δV = 2 ms^1 and δV = 10 ms^{-1}.

Weighting with the semimajor axis distribution of the dormant Halley-type comet populations (Levison *et al.*, 2002), the overall mean ratio of lifetimes is given by $L_m/L_c \sim 0.15$ (Fig. 5.2). Thus if the characteristic width of a meteor stream is ~0.1 AU, the number of strong Halley-type meteor streams intersecting the Earth's orbit according to Eq. (5.7) is ~15.

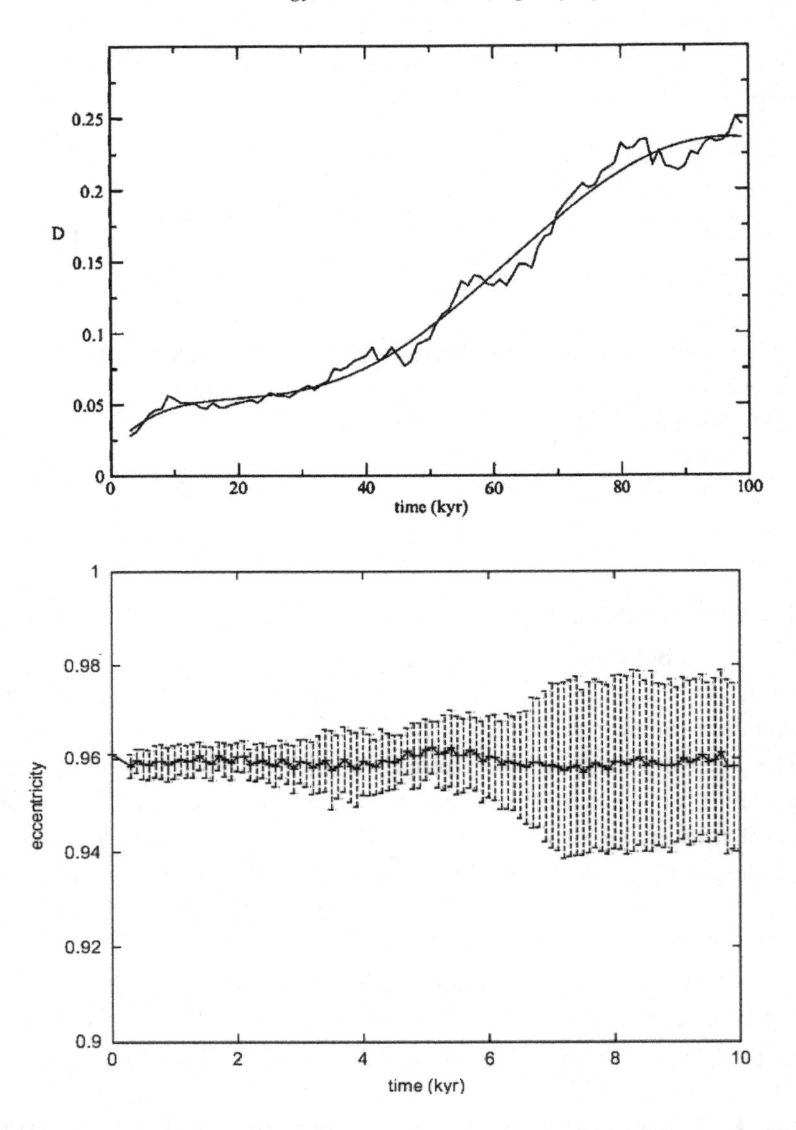

Fig. 5.1 Evolution of a meteor stream. The orbital evolution of 27 fragments of a Halley-type comet assumed to have disintegrated at perihelion. A standard similarity function (the *D*-criterion) is computed for each of the 27 orbits taken in pairs, and the average is plotted. The orbits gradually diverge under the influences of the planets. Characteristically a meteor stream will be recognised as such when $D \leq 0.2$ and so in this case the disintegrated comet (initial $q = 1.0$, $e = 0.96$, $I = 152^{\circ}$) would yield annual meteor showers over the full 10^5 yr of the integration.

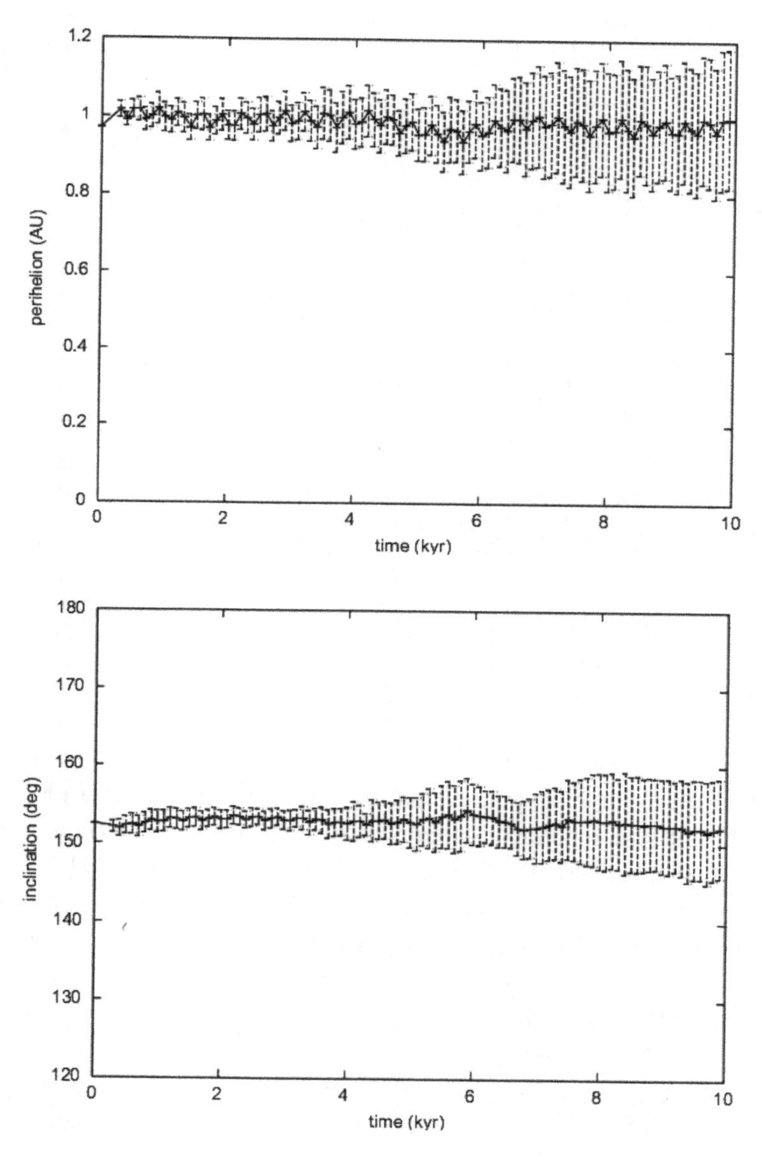

Fig. 5.1 (Continued).

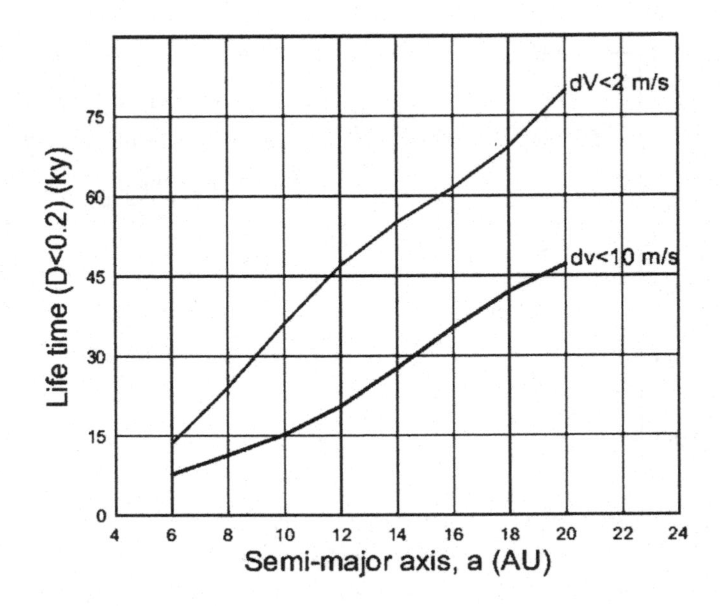

Fig. 5.2 Lifetimes of meteor streams with initial eccentricities in the range $0.96 \leq e \leq 0.98$, averaged and smoothed. 'Lifetime' is here defined as in the text, with $Dcrit = 0.2$. Lifetimes are shown for initial velocity dispersions, at perihelion, of $\delta V \leq 2$ ms^{-1} (upper curve) and ≤ 10 ms^{-1} (lower).

A meteor stream may be observed as more than one annual shower. Comet Halley, for example, yields the Orionids and η Aquarids. The somewhat diffuse meteoroid stream associated with comet Halley has a mass estimated at $\sim 5\times10^{17}$ g (McIntosh and Hajduk, 1983). The 30,000 Earth-crossing comets which are supposed to disintegrate completely have mean mass of this order or greater. One would then expect to observe streams of Orionid or η Aquarid strength (hourly visual rate ~20), unconnected with any visible comet, at weekly to monthly intervals. Such a situation is of course not observed.

Detection of the debris might be avoided if the comets disintegrate into sub-micron particles that do not contribute to zodiacal light. Entering at 50–70 kms^{-1} corresponding to Halley-type orbits, these would ablate in the high ionosphere and escape detection by both visual observations and radar. It would be necessary for 99% of comets in short-period orbits to be reduced to particles «0.1 mm in diameter within a few perihelion passages, a requirement which hardly seems plausible, and for which there is no independent evidence.

The dust concentration expected from the rapid disintegration of Halley-type comets has not been detected *in situ* by long-range spacecraft. The *Pioneer 10* dust instrument measured impacts out to 18 AU from the Sun before it eventually failed on May 1980, eight years after its launch. The orbital evolution of dust particles released from both retrograde and prograde Halley-type comets has been followed numerically for times much in excess of the survival time of meteor streams (Liou *et al.*, 1999). It was found that, for heliocentric distances r ranging from 1 AU to over 20 AU, the space density of the dust varied as $r^{-\gamma}$ with γ in the range 1.5 to 1.8. However, the *Pioneer 10* and *11* spacecraft found a constant dust density in the range 1–18 AU, in contradiction to the expected decline of $\sim 10^2$.

Modelling the total dust flux to include contributions from the Edgeworth–Kuiper belt objects and Jupiter-family comets, Landgraf *et al.* (2002) find that the Halley-type comets supply $\sim 3\times 10^5$ g s^{-1} of the observed dust. This amounts to $\sim 10^{15}$ g/century. However according to Bailey and Emel'yanenko (1998), comets brighter than absolute magnitude $H = 7$ are injected into Halley-type orbits an average of once every century. This corresponds to a diameter ~16 km which, for the density of water, yields a mass $\sim 1.5\times 10^{19}$ g. If all comets entered the Halley system with at least this mass, equilibrium would require 10^4 times more dust than detected by the *Pioneer* spacecraft. For a mean mass $\sim 10^{20}$ g the discrepancy becomes 10^5.

There is therefore a paradox: assuming a steady-state, the case for a large discrepancy rests only on Newtonian dynamics, the principle of conservation of mass, and the rate of influx of long-period comets, which is known to within a factor of a few (*cf.* Emel'yanenko and Bailey, 1998; Hughes, 2001).

To sum up, we should expect to see thousands of Halley-type comets in the sky, along with their decay products — either dormant bodies or annual meteor showers, but we do not. If we try to resolve the paradox by postulating wholesale disintegration into Tunguska-sized bodies, the Earth should be battered every few weeks by such bodies, and it is not. If we argue for wholesale disintegration into dust, we should see this in the form of strong annual meteor showers along with a bright, near-spherical zodiacal light. But all these entities are either absent or under-represented by factors of 10^2 or 10^3; nor have *in situ* measurements by the *Pioneer* spacecraft detected the anticipated dust flux.

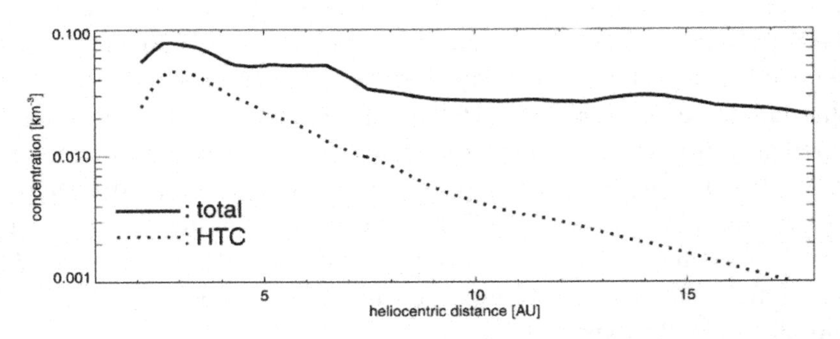

Fig. 5.3 The distribution of interplanetary dust in the outer solar system from *Pioneer* spacecraft (solid line). The relative contributions from the major sources (the Halley-type comet system; Jupiter-family comets; Edgeworth–Kuiper belt objects) are obtained from dynamical considerations and may be uniquely combined to fit the measured absolute flux. The best fit for the Halley-type system is shown as a dotted line and corresponds to a production rate of 0.3 tons/sec (after Landgraf *et al.*, 2002). This is at least 10^4 times too small for the mass balance problem to be explicable by the rapid disintegration of Halley-type comets into dust.

There is no good reason to assume detailed balance between the long period and Halley populations. However to account for the discrepancy the imbalance must be extremely large, of order 400–1600. Surges of this magnitude may occur at intervals $\sim 10^8$ yr when a star penetrates the hypothetical dense inner Oort cloud similar to the discussion in Chapter 4. However, comets in such a shower would have aphelia strongly concentrated around the region of sky where the perturbing star made its closest approach (Fernández and Ip, 1987), and this is not observed. In addition, the equilibration time between the long-period and Halley populations is only ~0.1 My and so the hypothesis would require us to live in a very special epoch, when a shower was underway but had not yet populated the Halley-type system.

5.1.2 Dark comets

Emel'yanenko and Bailey (1998), who first drew attention to the paradox, proposed that the Halley-type comets develop dark mantles. With the sublimation of ices and the collapse of dust back onto their surfaces, comets do indeed choke off their own outgassing activity and may become dormant or asteroidal in appearance; several comets have been observed to undergo this transition. The surfaces of comets are extremely dark, those which have been measured generally having albedos $p \sim 0.02$–0.05. Nevertheless such is the expected number of Halley-type comets, going down to say 2.4 km diameter, that significant numbers should by now have been detected by the *Spaceguard* surveys during close passages to the Earth. Assuming a mean albedo $p = 0.04$ for the dormant comets, Levison *et al.* (2002) estimated that about 400 such bodies should have been detected, but at epoch January 2002 only nine were known.

Photo 5.1 The *Giotto* spacecraft image of comet Halley in 1986 revealed the first great surprise to cometary scientists who had been wedded to a 'snowball comet' paradigm for several decades. This was the first time when the nucleus of a comet was clearly resolved. It turned out to be furthest in appearance to a snow field — it was described at the time as 'darker than the darkest coal'! Its albedo was measured at 0.03. Since 1986 even darker comets have been observed.

Instead of crumbling to meteoric dust, it is conceivable that the disintegration of the Halley-type comets would proceed to ~0.1 km 'Tunguska-sized' bodies, comparable with the dimensions of the smaller Kreutz sungrazers, or with the fragments such as were observed to split from comet Hyakutake (Desvoivres *et al.*, 2000), or for the comet Shoemaker Levy 9. Consider an isotropic distribution of N bodies of mean period P years, near-parabolic orbits and perihelia <1 AU. The mean collision probability of the Earth with such bodies, per perihelion passage, is $\sim 3 \pm 1\times10^{-9}$ (Steel, 1993). Thus with N/P perihelion passages per annum, collisions with the Earth occur at mean intervals δt yr given by

$$\delta t\left(\frac{N}{P}\right)3\times10^{-9}\sim 1 \tag{5.10}$$

That is,

$$\delta t\sim 3.3\times10^{8}\frac{P}{N} \tag{5.11}$$

to within about 30%. With the mass distribution (5.1) and α = 1.4, the mean mass of such an Earth-crosser is $\sim 1.1\times10^{20}$ g as we found from Eq. (5.4). Of the 3000 Halley-type comets with $q < 1.5$ AU, about 2000 are Earth-crossing ($q < 1$ AU). If all Halley-type comets break into 10^{12} g objects comparable to the mass of the Tunguska impactor, then one expects $\sim 3.3\times10^{12}$ fragments which, with $P \sim 60$ yr, would yield a Tunguska-like impact on Earth every few days!

5.1.3 Super-dark comets

A possible solution to the paradox is that the surfaces of inert comets become extremely dark. It is likely that the final stages of the accumulation of comets involved the mopping up of debris from the molecular cloud or protoplanetary disc. This accreted material was dominated by elongated refractory interstellar organic grains ~10^{-5} cm across, possessing volatile mantles (Bailey *et al.*, 1990; Wickramasinghe, 1974). Although silicate dust is also doubtless present, estimates of their mass fraction relative to the organic component could be as small as 15% (Crovisier *et al.*, 1997; Wickramasinghe and Hoyle, 1999). This is also consistent with mass spectra of interstellar dust obtained on instruments aboard the NASA spacecraft *Stardust* which showed an overwhelming dominance of heteroaromatic polymers and no evidence of minerals (Krueger *et al.*, 2004). On the grain model we consider, sublimation of volatiles would build up into a vacuous fairy-castle or aerogel structure. It is possible to envisage many metres of such a structure developing with little or no compaction, except perhaps through the agency of collisions with meteoroids. It is of interest in this connection that particles of probable cometary origin entering the ionosphere and stratosphere have a fluffy, porous structure, with vacuum filling factors ~0.75–0.95 (Rietmeijer, 2002; Wickramasinghe *et al.*, 2003). A tarry medium of refractive index $m_1 = n_1 - ik_1$, within which are distributed vacuum spheres of refractive index 1 occupying a fraction f of the total volume of material, behaves as one of complex refractive index m given by

$$m^2 = m_1^2 \left[1 + \frac{3f\left(1 - m_1^2\right) / \left(1 + 2m_1^2\right)}{1 - f\left(1 - m_1^2\right) / \left(1 + 2m_1^2\right)} \right] \tag{5.12}$$

where $m = n - ik$, n, k being the real and imaginary refractive indices respectively (Böhren and Wickramasinghe, 1977). For normal incidence the reflectivity R of a slab of this material is given by the formula from classical electromagnetic theory (Abraham and Becker, 1950)

$$R = \frac{(n-1)^2 + k^2}{(n+1)^2 + k^2} \tag{5.13}$$

For a prescribed tarry matrix and for given values of the vacuum volume fraction f we can readily solve for m from Eq. (5.4), and thus compute R. We take n = 1.45, and k = 0–0.1 at optical wavelengths, as being representative of organic polymers (e.g. kerogen). The results of numerical calculations are given in Fig. 5.4.

It is evident that very low values of R can be achieved with vacuum fractions in line with existing estimates for loose aggregations of interstellar dust grains: the bulk porosities of interplanetary dust particles are ~0.75, and those of Type III fireballs of cometary origin are ~0.95 (Rietmeijer, 2002). The fading function required by the observations may thus be achieved when a comet's surface, comprising a loose aggregate of interstellar dust grains, loses ice between the interstices. An active periodic comet loses gas and dust primarily through active areas on its surface. Some of this falls globally back onto the cometary surface (Wallis and Al-Mufti, 1996). In due course, with depletion of volatiles, the comet loses its coma and becomes dormant. The ice remaining in the volume between the surface grains sublimates, without replenishment, the cometary surface becomes an aerogel, and the albedo rapidly falls to very low values (Fig. 5.4).

With this model, it has been found experimentally that the thickness of the crust on the surface of a comet increases proportionately to the square root of the insolation time, while the gas production rate proceeds in inverse proportion to the thickness of the crust (Ibadinov *et al.*, 1991). Laboratory and numerical work (Ibadinov, 1993 and references therein) shows that the rate of fading of short-period comets, and its dependence on perihelion distance, are well reproduced by a nucleus of graphite particles embedded in water ice, with 80% porosity and thermal conductivity 0.05 $W^{-1}K^{-1}$. In this case, for a comet in a Halley-type orbit, the crust grows at ~5 cm yr^{-1}. The pressure of the escaping water vapour is insufficient to break such a crust (*loc. cit.* Kührt and Keller, 1996).

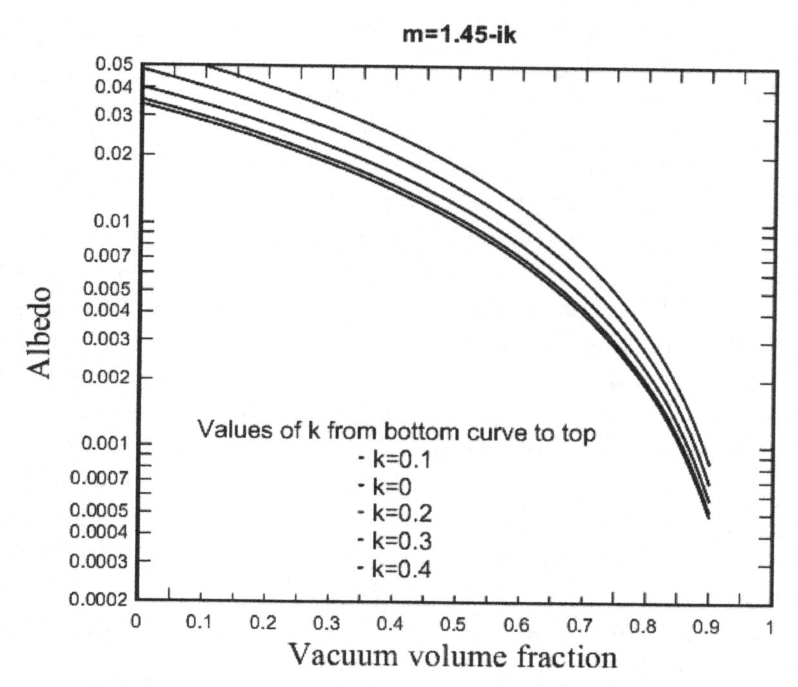

Fig. 5.4 The reflectivity of an aggregate of organic particles 10^{-5} cm in diameter as a function of porosity. For comparison, the nuclei of active comets have albedos $p \sim 0.02$–0.04, while dark spots on comet Borrelly have $p \sim 0.008$ (Nelson *et al.*, 2004). Brownlee particles and particles collected in stratospheric collections by Narlikar *et al.* (2003) are of probable cometary origin and have porosity of 75% or more.

The volume of space out to which a body of albedo p is detectable varies as $p^{3/2}$ and so, if there is an expectation that 400 dormant comets with canonical $p = 0.04$ should by now have been detected, the actual number of detections when $p = 0.002$ say is ~5 (*cf.* 9 detections to date). Thus the expected albedos of the carbonaceous aerogels are consistent with the presence of a large dark Halley population. The nearest known extraterrestrial albedos of this order are the dark spots on comet Borrelly, which have $p \sim 0.008$ (Nelson *et al.*, 2004). If the predicted population of dark Halleys had this albedo, perhaps about forty would by now have been discovered. There is clearly a strong selection effect against the discovery of astronomical objects in the solar system with albedos $p \ll 0.008$ (Jewitt and Fernández, 2001).

5.2 The Impact Hazard and the Panspermia Connection

The above solution to the 'missing comet' problem leads to the supposition that there exists a large population of extremely dark comets in Earth-crossing orbits. They are undetectable with current NEO search programmes but are impact hazards nonetheless.

Levison *et al.* (2002) computed that, without disintegration, there would be a population of $N \sim 3\times10^4$ dormant Halley-type comets with diameters $D > 2.4$ km and perihelia $q \lesssim 1$ AU. For $\overline{P} \sim 60$ yr and mean impact speed ~60 kms^{-1}, then from Eq. (5.11) the mean interval between impacts of such bodies is ~0.67 My, with impact energies $\gtrsim 1.5\times10^6$ Mt (Jeffers *et al.*, 2001). This leads to the conclusion that impacts of at least 1.5×10^7 Mt energy are expected at mean intervals »2.3 My. Averaged over geological time, this rate will be substantially less, since we appear to be in an impact episode now (Chapter 4), and there is broad agreement with the long-term average. But these rates are well in excess of those expected from the NEO system currently being mapped out (mainly S-type asteroids: Morbidelli *et al.*, 2002). Hughes (2003) has argued that there is no room for a significant cometary contribution, active or dormant, on the grounds that the impact rate from the near-Earth asteroid population is adequate to produce the known rate at which terrestrial craters are produced. This argument depends on a scaling relation between the diameter of the impactor and the crater which it forms, which is uncertain to order of magnitude. Rickman *et al.* (2001), on the other hand, find that comets yield a large, perhaps dominant, contribution to km-sized impactors. They estimate that a terrestrial impact rate of about one Jupiter-family comet (active or dormant) per My. The expected impact rate is also significantly higher than has been inferred from lunar cratering data (Neukum and Ivanov, 1994).

Current detection and deflection strategies involve the assumption that decades or centuries of warning will be available following the discovery of a threatening asteroid. However if the major impact hazard indeed comes from this essentially undetectable population, the warning time of an impact is likely to be at most a few days. A typical Halley-type dormant comet spends 99% of its time beyond the orbit of Mars and so a full mapping of this population is beyond current technology.

If the Halley-type population is derived in large part by capture of comets from the long-period system (Bailey and Emel'yanenko, 1998), then perturbations of the Oort cloud (as discussed in Chapter 4) may yield an upsurge in the dark Halley population, and ultimately in the flux of impactors on the Earth. As we have pointed out earlier, the Oort cloud is demonstrably sensitive to Galactic perturbers of various sorts — stars, nebulae and tides (Byl 1986; Napier and Staniucha, 1982 etc.). Nurmi *et al.* (2001) confirm that the flux of comets from the Oort cloud, and hence the impact rate, may fluctuate by an order of magnitude arising from the motion of the Sun with respect to the Galactic midplane. Since we are at present passing through the plane of the Galaxy, it is expected that the current impact rate is several times higher than that deduced from the lunar cratering record, which is time-averaged over one or two Gy.

The significance for the panspermia thesis is that the dark Halleys are a present-day link in the chain between the Galactic disturbances discussed in Chapter 4, huge impacts that dislodge microbiota from the Earth. The chief perturbers — giant molecular clouds — are concentrated in the spiral arms of the Galaxy, where of course star formation is proceeding. The idea that passages of the Sun through the spiral arms of the Galaxy might induce terrestrial disturbances has been discussed by a number of authors (e.g. McCrea, 1975). Napier and Clube (1979) specifically proposed that bombardment episodes might occur during such passages, leading to mass extinctions such as those which took place at the end of the Cretaceous.

Whether we can identify specific impact episodes with passages through spiral arms is another matter. The uncertainties are large. Leitch and Vasisht (1998) identified two great mass extinctions, the Cretaceous–Tertiary (~65 My BP) and end Permian (~225 My BP), with Sagittarius–Carina and Scutum–Crux arm crossings respectively. Gies and Heisel (2005), on the other hand, find the mid-points of recent spiral arm crossings at ~80 and ~156 My BP. Svensmark (2007) modelled the motion of the Sun in relation to the spiral arm pattern using a model-dependent hypothesis which has the Earth's past temperature as a proxy for encounters with spiral arms and found that the solar system passed through the Sagittarius–Carina arm ~34 My ago, and the Scutum–Crux arm ~142 My ago. Both these dates coincide with exceptionally strong bombardment episodes (Table 4.1). It seems that, at present, uncertainties in both the modelling of spiral arm kinematics and the strong incompleteness of the impact crater record preclude a secure identification of impact episodes or mass extinctions with specific spiral arm crossings.

On our model these bombardment episodes are in the main cometary, arising from disturbances of the Oort cloud as the solar system passes through a variable galactic environment. Along with mass extinctions, comet bombardment episodes (such as that at the K–T boundary) would directly contribute to panspermia by transferring genetic material from the Earth to nearby nascent planetary systems, and likewise in the opposite direction. Such material could go all the way from DNA fragments to complete plant seeds within life-bearing rocks à la Kelvin (Tepfer and Leach, 2006; Napier, 2004; Wallis and Wickramasinghe, 2004). It is likely, however, that exchange of life proceeds mainly through viable microorganisms. It is also conceivable that new genera are introduced via impacting comet material. In this connection, the sudden appearance of diatoms in the fossil record at precisely the time of the K–T impact is worthy of note (Hoover *et al.*, 1986).

Chapter 6

Expulsion of Microbes from the Solar System

6.1 Introduction

Throughout geological history a significant exchange of boulders between Earth, Mars and the Moon has taken place. Exchange of material between the Earth and Mars would have been most common between 4.6 and 3.8 billion years ago during the first 800 million years of the solar system's existence, when major impacts with asteroids and comets were frequent, the so-called Hadean epoch which was referred to in an earlier chapter.

Between thirty and forty meteorites have to date been identified as originating from the planet Mars, the most famous of which is meteorite ALH84001 (Hamilton, 2005). Identifications of their origin are based on the composition of trapped gases compared with the known abundances of these gases in the Martian atmosphere. Studies of ALH84001 (Weiss *et al.*, 2000) showed that the temperature of the rock never exceeded 41°C. This leads to the possibility that any bacteria lodged within the meteorite would retain their viability, hence its relevance to panspermia.

In this chapter we explore possible mechanisms for the exchange of life-bearing material between planets and also address the all-important question of survival of microorganisms during transit.

6.2 Expectations from Impact Cratering Mechanisms

The Earth accumulates an average of 20,000 tons of extraterrestrial material every year (Love and Brownlee, 1993). Most of this material enters Earth's upper atmosphere as small particles, but more rarely larger objects also strike Earth. These objects are rocky or metallic fragments of asteroids or cometary bolides, large and solid enough to survive passage through Earth's atmosphere.

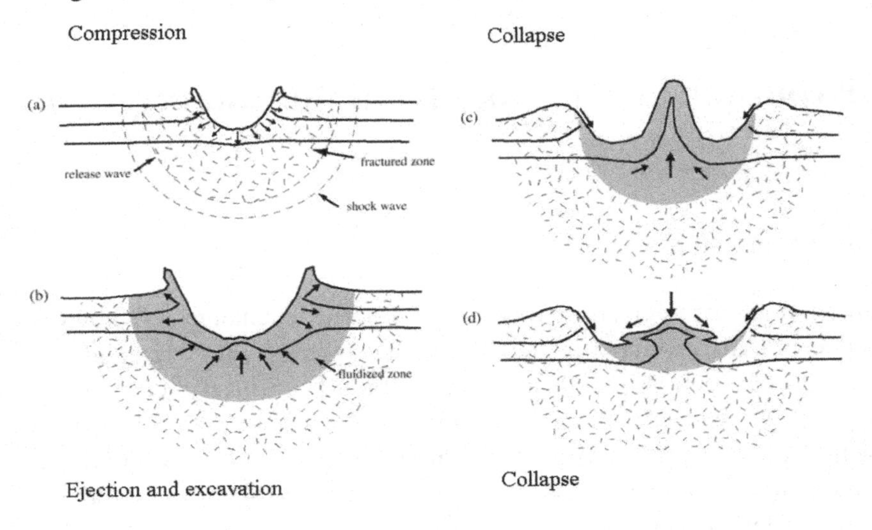

Fig. 6.1 Stages in the formation of an impact crater.

The formation of impact craters has been described by Melosh (1988, 1989). A hypervelocity impact takes place when a cosmic projectile penetrates the Earth's atmosphere with little or no deceleration from its original velocity. The process by which the crater is formed and modified can be described in several stages. Figure 6.1 shows a time sequence of events following an impact.

The nature of the crater that ultimately results from an impact depends on the size and velocity of the impactor. Two broad categories of craters can be identified, particularly from images of craters on the Moon which is geologically inactive:

(a) Simple craters with rim diameters typically less than 10 km showing bowl-like depressions and a well elevated rim, and

(b) Complex craters with rim diameters of 15–50 km showing slumped rims and a central uplift.

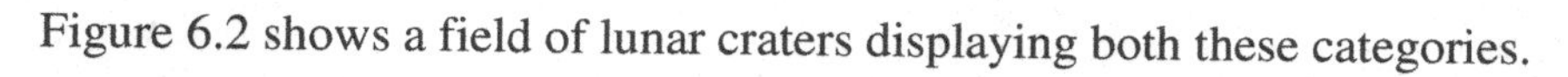

Figure 6.2 shows a field of lunar craters displaying both these categories.

Fig. 6.2 Showing a field of lunar craters (Courtesy NASA).

The minimum impact velocity of a collision with a comet or asteroid whose perihelion is close to 1 AU with Earth is 13.5 km/s. Velocities of impacting cometary bolides could be higher: for Earth-crossing bolides in Halley-family orbits we estimated average impact speeds of ~55 km/s (*cf.* Chapter 4).

We consider here the specific processes discussed in relation to cometary bolides. The immense kinetic energy of the impactor is transferred to the target rocks via shock waves at the contact or compression stage (O'Keefe and Ahrens, 1993). Energy loss occurs as the radius of the shock front increases, and as the target rocks are heated and deformed. It lasts perhaps only a few seconds, until the shock wave reflected back into the impactor is dissipated as heat. This leads to the almost complete melting and vaporisation of the impactor.

Then follows the excavation stage in which a transient crater is formed. Shock waves which surround the projectile expand through the target rock. As they travel upwards the resulting release waves are reflected downwards. Once the transient crater has reached its maximum size this stage is concluded and the crater is then modified under the influence of gravity and rock mechanics.

In the near-surface region the target rock is fractured and shattered, and some of the initial shock-wave energy is converted into kinetic energy. The rock involved is accelerated outward in the form of individual fragments that were not significantly heated, travelling at velocities that can exceed the escape speed from Earth (Melosh, 1988).

In this way, Melosh has described how impacts may throw meteorite-sized fragments of rock into space, a fraction of which could be transferred to other planets in the solar system and beyond.

6.3 Mechanisms for Ejection and Fragmentation of Boulders

Lunar and martian meteorites provide evidence of actual material that has undergone this process, while the Chicxulub crater event 65 My ago provides evidence of the collisional ejection process. Wallis and Wickramasinghe (1995) showed that ejecta from such a terrestrial event readily reach Mars. Mileikowski *et al.* (2000) discussed this transfer process, requiring individual metre-sized rocks to be ejected at very high speeds through an atmosphere of density ~10^4 kg m^{-2}. Wallis and Wickramasinghe (2004) discuss an alternative scenario. For an impact of a large comet, for example the Chicxulub impactor, the sheer bulk of ejecta exceeds the atmospheric mass by a large factor. Rocks and debris can reach escape speeds by being swept up with the high speed vapour and particles. Such a massive plume is produced by the collision that it essentially blows a hole through the atmosphere (Melosh, 1989). The implication is that rock fragments and debris down to millimetre sizes or smaller can be ejected along with larger rocks. This material could carry away entire micro-ecologies of life into space.

A proportion of boulders ejected from the topmost layers of an impact site may reach escape velocity relatively unshocked and so are capable of bearing live microorganisms (Section 4.1.2). Fragmentation and erosion of the boulders occur within the dense zodiacal cloud over 10^4 years (Napier, 2004). The disintegrated particles are ejected from the solar system as beta-meteoroids (particles for which the repulsion due to sunlight exceeds the attraction due to gravity) and are then reincorporated into protoplanetary systems during passages through dense molecular clouds, particularly at times when the solar system crosses the mid-plane of the Galaxy.

6.4 ß-Meteoroids

A requirement in cometary panspermia theories is for small (micrometre-sized) grains (including bacteria) to be ejected entirely out of the solar system and a fraction of these to be trapped and incorporated within an embryonic planetary system. Small grains generated in the inner solar system and leaving the solar system on hyperbolic orbits due to the dominating effect of the radiation pressure force were identified by Zook and Berg (1975).

The ratio of radiation pressure force to solar gravity is designated β. Particles with $\beta > 1$ are called β-meteoroids. The existence of these particles was confirmed by Hiten satellite measurements (Igenbergs *et al.*, 1991).

The motion of such particles follows well-attested principles of Newtonian orbit theory. For an attractive force of μr^2 per unit mass, the equations of motion integrate to give

$$\frac{d^2 u}{d\theta^2} + u = \frac{\mu}{h^2} \tag{6.1}$$

where $u = 1/r$, h is angular momentum and θ is the angle in polar coordinates. Integrating Eq. (6.1) gives

$$u + \frac{\mu}{h^2} = A\cos(\theta + \varepsilon) \tag{6.2}$$

If a particle is projected from a distance $c = 1$ AU with velocity $\sqrt{2\mu / c}$ at angle α to the radius vector then by differentiating Eq. (6.2) we obtain

$$-\frac{1}{r^2}\frac{dr}{d\theta} = -A\sin(\theta + \varepsilon) \tag{6.3}$$

It can be shown that initially, for θ=0,

$$r\frac{d\theta}{dr} = \tan\alpha \tag{6.4}$$

which gives

$$\left(\frac{r}{c}\right)^{-1} = -\frac{\beta}{2\sin^2\alpha} + \frac{\cot\alpha}{\sin\varepsilon}\cos(\theta + \varepsilon) \tag{6.5}$$

For given values of β and α we can therefore compute the trajectory of a particle as it is ejected hyperbolically. This is done in Sections 6.7–6.9.

6.5 Protective Shielding in Small β-Meteoroids

The effects of ultraviolet light and low energy galactic cosmic rays have continued to be discussed as a problem for the survival of bacterial particles in interstellar transit. Cometary panspermia theories require a fraction of bacterial particles to be available in viable form in order to seed embryonic cometary/planetary systems forming elsewhere in the galaxy (Hoyle and Wickramasinghe, 1981).

Here we shall show that the hazard of galactic cosmic rays is not sufficient to reduce viability of particles to the extent of compromising panspermia. Only the minutest survival fraction, possibly less than 1 in

10^{24}, is required to maintain a positive feedback loop for cometary panspermia to work. The best prospect for retaining viability is for small clumps (colonies) of bacteria travelling together as self-shielded integral units. In such cases interior organisms would be well protected from damaging ultraviolet radiation. Bacterial clumps of this kind, with diameters in the range 1–10 μm, have been recovered from the stratosphere up to heights of 41 km and have been provisionally interpreted as being of cometary origin (Harris *et al.*, 2002).

The question arises as to the confinement or otherwise of these clumps within the solar system. Under certain circumstances can desiccated bacteria, individually or in small clumps, be accelerated to the outermost regions of the solar system and beyond? And under what conditions do they survive long interstellar journeys before becoming incorporated in a new generation of comets? We attempt to address these issues in the rest of this chapter.

6.6 Carbonisation of the Surface Layers of Grains

A small clump of desiccated bacteria after prolonged exposure to solar ultraviolet radiation at ~1 AU, would inevitably become charred/graphitised at the surface. Taking the interior bacterial material to have an estimated volume filling factor of ~30% (60–70% of the volume of a vegetative cell is water), its average refractive index will be $< n > \cong 1.0 + 0.3(1.5 - 1) = 1.15$, assuming n = 1.5 for the hydrated bacterial material (Hoyle and Wickramasinghe, 1979).

Solar ultraviolet photons acting on the surface would eventually lead to the formation of a thin layer of reduced carbon, but the growth of this layer would be self-limiting. A graphite sphere of radius 0.02 μm has a peak extinction efficiency $Q_{ext} \approx 3.348$ at the ultraviolet wavelength of 2175A giving a mass extinction coefficient of 5.7×10^5 cm^2 g^{-1}, assuming a bulk graphite density of s = 2.2 g cm^{-3} (Wickramasinghe, 1973). A closely similar mass extinction coefficient will also be relevant for a thin layer of graphite (Wickramasinghe, 1967). A layer of thickness t comprised of such material will thus produce an optical depth

$$\tau = 5.7 \times 10^5 \times 2.2\, t \tag{6.6}$$

A value $\tau = 3$ will be achieved for a graphite thickness of $t = 0.024$ μm. Therefore it is unlikely that the overlying graphite layer will grow to much more than ~0.03 μm before the ingress of ultraviolet photons is checked. In this case genetic material residing in the interior would be well protected from damaging solar ultraviolet light.

6.7 Radiation Pressure Effects

We next model the effects of radiation pressure on a spherical clump of hollow bacterial grains of radius a surrounded by varying thicknesses t of graphite mantle. We assume a refractive index $m = 1.15$ consistent with the discussion in Section 6.6. (Here we ignore the effect of a small absorption coefficient that may apply within the cells.) For the overlying graphite mantle we adopt the wavelength dependent values of n, k from the laboratory data of Taft and Phillipp (1965) (see tabulations in Hoyle and Wickramasinghe, 1991).

Each particle is now regarded as an idealised concentric core-mantle grain. The dielectric core radius is defined as $r_c = a$, and the outer mantle radius is $r_m = a + t$. For given values of (a, t) we can calculate at any wavelength λ the optical efficiencies $Q_{ext}(\lambda)$, $Q_{sca}(\lambda)$, and the phase parameter $\langle\cos\theta\rangle$ from the Güttler formulae for coated spheres (Wickramasinghe 1973).

The efficiency factor for radiation pressure $Q_{pr}(\lambda)$ is then given by

$$Q_{pr}(\lambda) = Q_{ext}(\lambda) - \langle\cos\theta\rangle Q_{sca}(\lambda) \tag{6.7}$$

For the purpose of evaluating the effect of solar radiation on grains we next compute an approximation to the ratio

$$\overline{Q}_{pr} = \frac{\int_0^\infty F_\lambda Q_{pr}(\lambda)\, d\lambda}{\int_0^\infty F_\lambda\, d\lambda} \tag{6.8}$$

for each of the cases under discussion.

Here $F_\lambda d\lambda$ is the relative energy flux of sunlight in the wavelength interval ($\lambda, \lambda + d\lambda$).

6.7.1 Ratio of radiation pressure to gravity

A bacterial grain (or grain clump) of external radius r_m located at a distance R from the centre of the Sun of radius $R_\odot$ ($R \gg R_\odot$) will experience a force directed radially outward due to radiation pressure of magnitude

$$P = \frac{L_\odot}{c} \pi r_m^2 \overline{Q}_{pr} \frac{1}{4\pi R^2} \tag{6.9}$$

where $L_\odot$ is the bolometric luminosity of the Sun and c is the speed of light. The oppositely directed gravitational attractive force to the Sun is

$$G = k \frac{m M_\odot}{R^2} \tag{6.10}$$

where $M_\odot$ is the solar mass, m is the grain mass and k is the universal gravitational constant. The mass m of the grain clump is given by

$$m = \frac{4}{3}\pi \left[(a+t)^3 - a^3\right] s + \frac{4}{3}\pi a^3 \rho \tag{6.11}$$

where s is the bulk density of graphite and ρ is the mean density of the hollow clump material. With s = 2.2 g cm^{-3}, ρ = 0.3 g cm^{-3} and $L_\odot/M_\odot$ = 3.83/1.99, Eqs. (6.9), (6.10) and (6.11) yield

$$\frac{P}{G} = 0.57 \frac{(a+t)^2}{2.2(a+t)^3 - 1.9a^3} \overline{Q}_{pr} \tag{6.12}$$

where a and t are in micrometres. Our computations of $\overline{Q}_{pr}$ described in Section 6.7 combined with Eq. (6.12) then gives P/G ratios for each of the grain models considered. The results are plotted in Fig. 6.3.

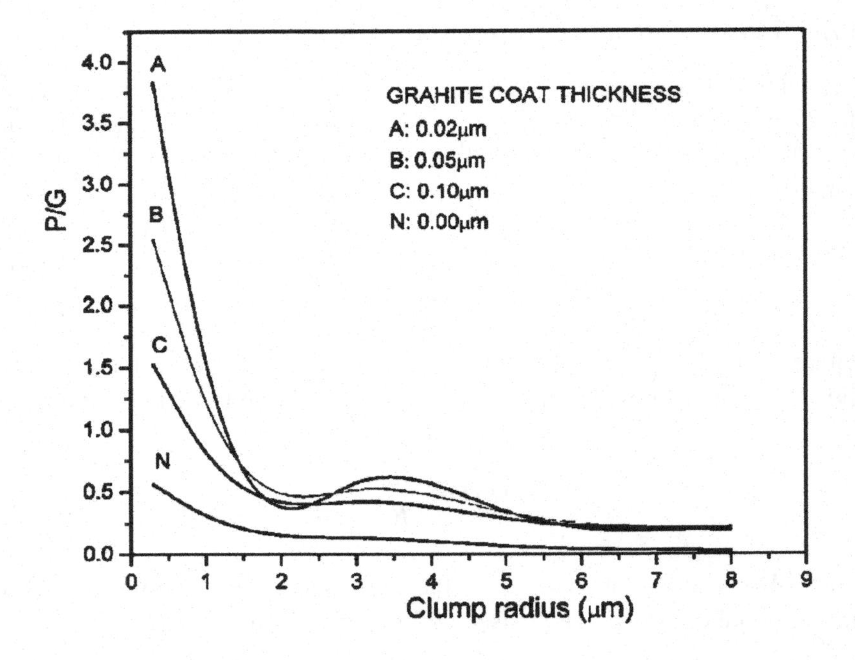

Fig. 6.3 The ratio of radiation pressure to gravity for bacterial grains/grain clumps of radius a with concentric mantles of graphite of thickness t.

6.7.2 Results and dynamical considerations

For particles without graphite coatings the curve marked (N) shows that the ratio *P/G* stays well below unity for all the cases we have considered. This means that such grains will remain gravitationally bound to the solar system and will not readily contribute to interstellar panspermia.

The curves A, B, C show, however, that this situation changes dramatically as the particles begin to acquire carbon coatings. For a graphite coating of thickness 0.02 μm *P/G* exceeds unity for bacterial spheres of diameters in the range 0.6–2.5 μm. The lower limit is appropriate to a single bacterium, or a few small bacteria, and the upper limit to small clumps of bacteria. The smaller particles and clumps once carbonised at the surface are seen to be well suited as candidates for interstellar panspermia. Once expelled from the solar system they could reach nearby protosolar nebulae in timescales of a few million years. Much larger clumps, on the other hand, as well as those with greater thicknesses of exterior graphite will, on the whole, tend to be gravitationally bound to the solar system.

For a grain satisfying $P/G > 1$, radial acceleration at a heliocentric distance R would be governed by the equation

$$\ddot{R} = \alpha \frac{kM_{\odot}}{R^2} \tag{6.13}$$

where $\alpha = P/G - 1$. (Gas drag at interplanetary densities of $\rho_H \approx 10^{-22}\, gcm^{-3}$ is negligible within the solar system.) Integration of Eq. (6.13) then leads to

$$\dot{R}^2 = -2\alpha kM_{\odot}\left(\frac{1}{R} - \frac{1}{R_0}\right) \tag{6.14}$$

where R_0 is the heliocentric distance at which radial acceleration of the grain is assumed to begin. As $R \to \infty$ the asymptotic velocity given by Eq. (6.14) is

$$V_{\infty} = \sqrt{\frac{2\alpha k M_{\odot}}{R_0}} \tag{6.15}$$

Taking $\alpha = 0.5$ (P/G = 1.5) as a typical case and R_0 = 1 AU, Eq. (6.15) gives $v_{\infty} \approx 3\times10^6$ cm/s. With this speed a bacterial clump would reach the outer edge of the Edgeworth–Kuiper belt, ~100 AU in ~15 years and a protosolar nebula in a perturbing molecular cloud (such as discussed in Chapter 4) at ~10 pc in ~0.3 My. Survivability during interstellar transit is therefore required for a modest timescale of well under a million years.

6.8 Surviving the Hazards of Galactic Cosmic Rays

Whilst a thin layer of graphitised carbon around a bacterium or clump of bacteria provides complete protection from ultraviolet light, exposure to galactic cosmic rays poses a more serious potential threat (Mileikowsky *et al.*, 2000). Radiation doses that limit viability appear to be critically dependent on bacterial species. Below is a table showing the percentage of spores of *Bacillus subtilis* surviving after naked exposure to space vacuum in the solar vicinity (Mileikowsky, 2000).

Table 6.1 Survival of spores of *B. subtilis* (data from Horneck, 1993, Horneck *et al.*, 1995).

Mission	Duration of vacuum exposure (days)	Survival fraction at end of exposure (%)
Spacelab 1	10	70
EURECA	327	25
LDEF	2107	1.4 ± 0.8

Within the solar system the radiation doses received by an outward moving bacterium depend critically on the phase of solar activity, being highest at times near the peak of the solar cycle. The average radiation dose received at 1 AU over a timescale comparable to the 11 yr period of the solar cycle is less than 10 Mrad, and so viability will not be expected to be drastically reduced. The LDEF experiment indeed shows a reduction to 1% for the extreme case of an unprotected, unfrozen (at ~1 AU) bacterial culture. Cryogenically-protected clumps of bacteria or bacterial spores are expected to fare significantly better. In a typical interstellar location a very much lower flux of ionising radiation would be delivered over astronomical timescales. There are indications that a higher tolerance would be expected in this case. We shall return to this point later in this chapter.

We next estimate theoretically the doses of ionising radiation intercepted by bacterial clumps in an unshielded interstellar cloud. To do this we compute the rate of deposition of ionisation energy E due to cosmic ray nuclei passing through a spherical grain of radius a (in microns) and density $s \approx 1$ g cm^{-3}:

$$\frac{dE}{dt} \cong Ja \ \mathrm{MeV\,cm^{-2}\,s^{-1}} \tag{6.16}$$

where J is given by

$$J = \left(\sum_i f_i Z_i^2\right) \int_{1MeV}^{\infty} F(E) \frac{1MeV}{E}\, dE \tag{6.17}$$

Here f_i denotes the fraction of cosmic ray nuclei with atomic number Z_i, and $F(E)dE$ is the flux of cosmic ray protons with energy in the range (E, E+dE) (Salpeter and Wickramasinghe, 1969). The value of J would vary from place to place in the interstellar medium, and is in general dominated by a low energy tail of the cosmic ray spectrum which is cut off here at E = 1 MeV.

For a spherical grain of radius a, Eq. (6.16) gives an energy dissipation rate into solid material

$$q = \frac{4\pi a^2}{4\pi a^3 s/3} Ja \quad \text{MeV g}^{-1}\,\text{s}^{-1}$$

which with $s = 1$ g cm^{-3} converts to

$$q \approx 1.5 \times 10^4\, Jr \text{ yr}^{-1} \tag{6.18}$$

remembering that a radiation dose of 1 rad corresponds to a deposition of ionisation energy of 100 erg g^{-1}.

The much-publicised adverse effects of ionising radiation in space can be shown to be flawed from a number of standpoints. A value of J in the range 0.01–0.1 cm^{-2} s^{-1} within interstellar clouds seems plausible in the light of available astronomical data (see, for example, arguments in Spitzer, 1978). A value J = 0.01 cm^{-2} s^{-1} thus delivers 15 Mrad in 10^5 yr.

Consider now the residual viability of a hypothetical microbial species that halves its viable fraction with a radiation dose of, say, 1.5 Mrad. Over the 0.3 My timescale for reaching the cosy protection of a protosolar nebula located ~10 pc away (see Section 6.9), and consequent inclusion in a new generation of comets, the integrated radiation dose received is 45 Mrad, leading to our hypothetical species being attenuated by a factor $(1/2)^{30} \approx 10^{-9}$. For a more radiation-susceptible microbial species that halves its viable population with a radiation dose of 750 krad, the corresponding viable fraction reaching the new protosolar nebula will be 10^{-18}. Even such an extreme rate of attenuation will not be necessarily lethal for panspermia.

To incorporate a million viable bacteria of this type in every comet condensing in the new system requires the accommodation of a total of 10^{24} itinerant bacteria, dead and alive. This would contribute only ~10^{11} g, which is less than one tenth of a millionth of the mass of a 10 km-sized comet. According to the present argument interstellar panspermia will be assured even with the most pessimistic of assumptions. But the most relevant experimental results could be even more encouraging.

Experiments done so far to determine viability as a function of total dose are most likely to be irrelevant for this purpose. Laboratory studies have invariably used very high fluxes of ionising radiation delivered in short pulses, at room temperature in the presence of air. Extrapolation from these to the astronomical case involves many uncertainties. In particular, most damage resulting from free-radical formation will be eliminated in an anaerobic environment, and low flux-long time intervals may not equate to short pulses of high-flux radiation in the laboratory case.

Similar conclusions follow from a re-evaluation of laboratory data by Horneck *et al.* (2002). They discuss experiments to determine the survival of *B. subtilis* spores subjected to encounters with simulated heavy nuclei in galactic cosmic rays. It has been possible to 'shoot' such nuclei to within 0.2 μm of a spore. The fraction of inactivated spores at various distances of the cosmic ray (CR) track are given as

$$b \leq 0.2\ \mu\text{m}: 73\ \%$$
$$0.2\ \mu\text{m} < b < 3.8\ \mu\text{m}: 15\text{–}30\ \%$$

These experiments show that even for a central hit by a high Z low energy CR particle a significant fraction would survive, and since the flux of such particles is low, a timescale of ~1 million years must elapse before a spore would be deactivated.

Citing other experiments, Horneck *et al.* (2002) point out that for spores embedded in a meteorite, the timescale for survival to a 10^{-4}% level (due to primary and secondary CR effects) at a depth of 2–3 m is ~ 25 My. The same surviving fraction is obtained for shielding at a depth of 1 m after 1 My, and for a naked spore the estimated timescale is ~0.6 My. Over timescales of 3×0.6 My, the survival fraction is 10^{-18}. A clump of bacterial spores travelling at 10 km/s would traverse a distance of 18 pc during this time, comfortably less than the mean distance between cometary/planetary systems. Of course during close approaches of molecular clouds the transit time between the solar system and an embryonic planetary system will be <10^4 years, over which much larger survival fractions are guaranteed.

There are strong indications that even the survival rates quoted above are a gross underestimate. They are based on experiments conducted at atmospheric pressure with normal levels of humidity (Lindahl, 1993).

There is evidence that bacterial endospores are resistant to inactivation by free-radical damage and chemical processes because their cytoplasm is partially mineralised and their DNA is stabilised (Nicholson *et al.*, 2000). Most significantly it has been found that bacterial spores (*genus Bacillus*) in the guts of a bee preserved in amber for 25–40 My could be revitalised (Cano and Borucki, 1995). Similarly, bacterial spores in 250 My old salt crystals have also been revitalised (Vreeland *et al.*, 2000). In both these studies the most stringent sterilisation techniques were used to avoid contamination of the samples, and the authors are confident that they have revived bacterial spores of great ages. With a natural background radioactivity (of rocks) of ~1 rad per year, we have here evidence of survival with doses of 25–250 Mrad of ionising radiation.

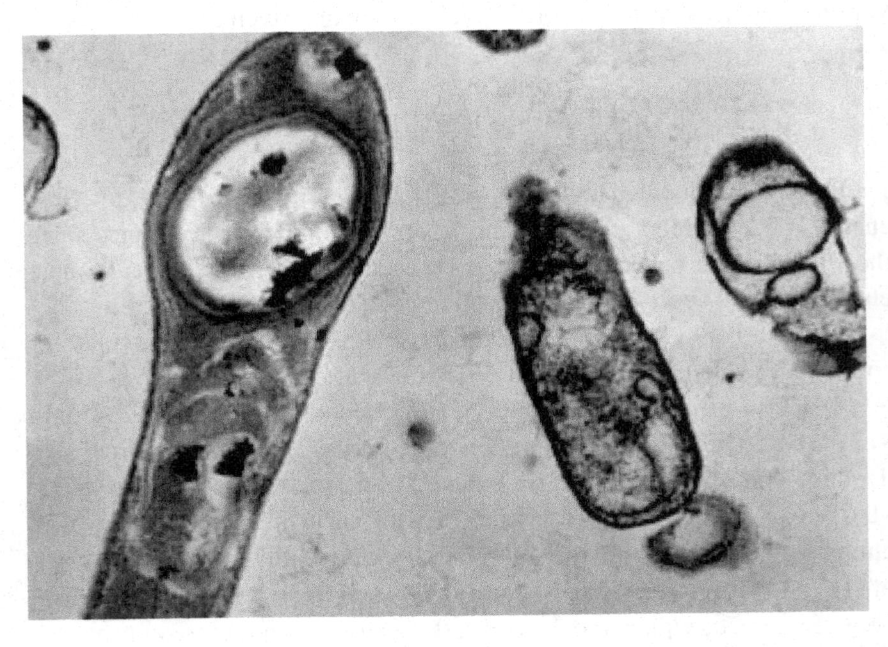

Fig. 6.4 Salt crystals from a New Mexico salt mine in which 250 My viable bacteria were recovered (Vreeland *et al.*, 2000).

More recently Bidle *et al.* (2007) have examined the most ancient terrestrial ice deposits in the Antarctic and found evidence of viable bacteria/bacterial genes in 8 My old ice. They estimate an average half-life of bacteria under such conditions of ~1.1 My. The importance of these results in relation to panspermia cannot be denied. To avoid this conclusion some scientists have used phylogenetic arguments. They point out that the revived organisms bear a close relationship to contemporary species, and so argue that they must be contaminants. However, panspermia theories permit re-introductions of the same organism (stored in comets) separated by millions of years, so no evolutionary divergence is required. Recent phylogenetic studies have indeed shown that 'horizontal' gene transfers across branches of Carl Woese's tree of life befuddle attempts to use the 'tree' as guiding principle or as an evolutionary clock (Woese and Fox, 1977). The phylogenetic modernity of the microorganisms found, as well as their survival over millions of years, consistently point to an extraterrestrial origin in our view.

6.9 How Comets Distribute Life

Comets play a role in panspermia in one of two ways:

(1) Collecting viable microorganisms (bacteria and archaea) from interstellar clouds during comet formation in the early solar system, amplifying them in radioactively heated comet interiors, and re-infecting the interstellar medium with a new generation of microbes.

(2) During comet impacts onto inhabited planets like the Earth, microorganisms in the uppermost layers (including products of local evolution) could be expelled within boulders that are subsequently ground into micron-sized dust via collisions with zodiacal dust.

In both cases micron-sized solid particles enter the interplanetary medium and a fraction which are β-meteoroids are expelled entirely from the solar system. This process is most efficient during encounters with molecular clouds over typical timescales of ~1 My. Clumps of dust-bearing microbes leave the solar system at speeds that enable them to reach distances of ~10 pc within a million years. The passing molecular cloud then encounters an extended 'biosphere' of the Earth, leading to transfers of terrestrial microbiota to protoplanetary nebulae within it. The microbial particles are stopped by gas drag within nascent planet/comet-forming clouds.

Option (1) requires that some fraction of freeze-dried bacteria and spores survive interstellar travel for typical timescales of 10^4–10^6 years. Several hazards have to be endured; these include freeze-drying, exposure to ultra-violet radiation and exposure to cosmic rays. Freeze-drying (if done slowly, as would be the case for planetary or cometary escape) poses no problem, and UV radiation whilst potentially damaging is easily shielded against by surface carbonisation of clusters (as we showed earlier), or by embedding within porous dust particles. The main loss of viability would ensue from cosmic ray exposure, the high Z (atomic weight), low energy particles being potentially the most damaging. It will be shown in Section 7.1 that only the minutest survival fraction of microorganisms is required in order to make cometary panspermia in option (1) work.

For 10μm-sized clumps of bacteria entering an atmosphere like the Earth's, it could be argued that frictional heating would not be enough to limit viability (Coulson, 2003; Coulson and Wickramasinghe, 2004). Individual bacteria or small clusters of bacteria would survive entry into a planetary atmosphere more easily in shallow-angle encounters. Incoming particles become satellites of the destination planet in the first instance, and subsequently spiral in.

6.10 Dispersal of Life by Impacts

For large boulders ejected during impacts on Earth (asteroid or comet impacts) transfer of embedded life proceeds along a different route. Impactors of diameters >1 km are most efficient. Considerable amounts of life-bearing rocks are thrown up at high velocities, some fraction reaching escape speeds. Ejecta leave the Earth and orbit around the Sun for timescales varying from several hundred thousand years to mega years. A fraction leaves the solar system, a small fraction of which reaches other planetary bodies. If conditions are appropriate, terrestrial life can seed an alien planet.

There are, however, several bottlenecks to cope with before viable transfers take place:

(1) Residual survival in the impact ejection from the source planet
(2) Residual survival in transit — enduring radiation environment of space
(3) Residual survival on landing at a destination (Earth-like) planet

Shock damage caused by the ejection process could be significant and lead to a reduction in viability of microbes embedded within rocks. Burchell *et al.* (2004) have attempted to quantify this reduction by deploying a gas gun to fire pellets containing microorganisms onto solid surfaces. Their results, which were surprising at first, showed survival rates of 10^{-4}–10^{-6} in several cases where shock pressures in excess of 30 GPa were applied. For *Bacillus subtilis* a survival fraction of 10^{-5} was found at a shock pressure of 78 GPa, with smaller yet finite survival fractions for higher values of the pressure. Stoffler, Horneck, Ott *et al.*, (2007) have independently confirmed high survival rates for *Bacillus subtilis* as well as for other microorganisms exposed to shock pressures of ~ 45 GPa. This is near the upper limit of shock pressures inferred in Martian meteorites and would be appropriate to most impact-driven ejection processes.

The terrestrial analogue of panspermia is the dispersal of plant seeds across the surface of the planet. With many millions of seeds produced by an individual plant only a single viable seed needs to find a fertile landing site to re-establish the species in a different location. The similarity to the logic of panspermia cannot be overlooked. Indeed Svante Arrhenius (1908) cited the long-term cryogenic survival properties of plant seeds as one of his main justifications for panspermia. Plant seeds have been known to retain viability in archaeological sites which are thousands of years old, and Tepfer and Leach (2006) have obtained further evidence for survival of seeds against the many hazards of space. With lithopanspermia the possibility does indeed arise that plant species could be translocated on a galactic scale, or at any rate that genetic information in the DNA of plants could be safely and securely carried within seeds in small rocks and boulders.

Chapter 7

Liquid Water in Comets

7.1 Introduction

Despite the fact that water-ice has been known to make up a large fraction of a comet's composition, liquid water in comets was considered impossible for a long time. The dominance of Whipple's icy conglomerate model with comets sublimating from a perpetually frozen material prevented any serious discussion of a phase transition from ice to liquid water (Whipple, 1950). The first discussion of liquid water in comets can be traced to Hoyle and Wickramasinghe (1978):

"Typically a cometary body would acquire an organic mantle about a kilometre thick. The organic mantles (comprised of interstellar molecules) would initially be hard frozen like the ices below them... Internal heating and liquefaction inevitably occurred, however, as a result of chemical reactions among the organic materials, triggered perhaps by collisions of the comets with smaller bodies. Such reactions could release up to ten times the energy needed to melt cometary ices at a depth of a few hundred metres below the cometary surface..."

A necessary condition for liquid water to be stable within a comet is that the ambient temperature and pressure exceed the corresponding triple point values, $T = 273$ K and $p = 6$ mb. For a static uniform sphere of radius r and density ρ the central pressure is

$$p = \frac{2\pi}{3} G\rho^2 r^2 \qquad (7.1)$$

Assuming $\rho \approx 1$ g cm^{-3} and setting $p > 6$ mb, Eq. (7.1) yields

$$r > 2\,\text{km} \tag{7.2}$$

This is clearly the minimum radius of a cometary body that can sustain a liquid water core.

The phase change from ice to liquid water could produce dramatic biological consequences in a comet. The smallest mass of viable microbial material entering a nutrient-rich watery medium could be vastly amplified on a very short timescale.

The generation time (doubling time) of a microoganism depends on type and species, nutrient concentration as well as temperature. A naturally-occurring population of bacteria in Siberian permafrost is known to have a doubling time ranging from 20 days at –10°C to 160 days at –20°C (Rivkina *et al.*, 2000). For a nutrient-poor cometary medium at a temperature close to 270 K let us assume the doubling time for an anaerobic bacterium to be ~1 yr (which is likely to be an upper limit).

If τ is the generation time of the bacterium, the biomass at a time t is given by

$$m \cong m_o 2^{t/\tau} \tag{7.3}$$

where m_0 is the intial biomass. Equation (7.3) can be re-written to give

$$t = \frac{\tau \ln\left(m/m_o\right)}{\ln(2)} = 1.44\tau \ln\left(\frac{m}{m_o}\right) \tag{7.4}$$

With $m_0 = 10^{-12}$ tonne (~10^6 bacteria) a million tonnes of biomass (representing <10^{-6} of a typical cometary mass) would be generated in a mere 60 years. Thus the smallest initial mass of viable bacteria/spores included within a newly-formed comet could be vastly amplified within its melted interior in an astonishingly short timescale. This confirms the importance of comets as an amplification site in relation to panspermia.

7.2 Primordial Melting

Melting of cometary ices might occur due to heat generated by radioactive elements. Two short-lived nuclides that can be considered in this context are ^{26}Al (half-life 0.74 My) (Wallis, 1980) and ^{60}Fe with a half-life 1.5 My. The energy released by radioactive decay of ^{26}Al to ^{26}Mg, is 1.48×10^{13} J/kg. With an interstellar isotope ratio of $^{26}Al/^{27}Al \approx 5\times10^{-5}$ and the known cosmic abundance ratios of C,N,O/Al we obtain a ^{26}Al mass fraction in pristine cometary material of $\sim3\times10^{-7}$ (MacPherson *et al.*, 1995; Diehl *et al.*, 1997). The total energy available from radioactive decay of ^{26}Al within a comet is thus $\sim4.4\times10^{6}$ J/kg. Recently the importance of ^{60}Fe in heating planetesimals has been discussed (Mostefaoui *et al.*, 2004) and similar considerations apply to comets. With relative abundance an order of magnitude higher in cometary material, and a higher energy yield per unit mass, the heat available from the decay of the two shorter-lived nuclides ^{26}Al and ^{60}Fe could total 3×10^{7} J/kg. This energy yield is over two orders of magnitude higher than the heat of fusion of water-ice ($\sim3.34\times10^{5}$J/kg).

The feasibility of primordial liquid water cores is thus firmly established but estimating the duration of a liquid phase requires more detailed calculations. Such calculations involve several poorly-defined parameters, including the thermal conductivity of the cometary regolith and the time lapse between the injection of radionuclides into the solar nebula and the formation of comets. Such uncertainties are alleviated, however, by the extreme shortness of the required timescale for bacterial replication (see Eq. 7.4).

Detailed model calculations of the thermal evolution of radiogenically-heated comets have been carried out by Wallis (1980), Yabushita (1993), Merk and Prialnik (2003) among others. A limiting factor for heating by short-lived nuclides (^{26}Al, half-life 0.74 My; ^{60}Fe half-life 1.5 My) is the time taken for the cometary bodies to accrete following the injection of radioactive nuclides into the solar nebula. If the process takes much longer than several million years, these radioactive heat sources would have become extinct. Uncertainties still exist in theories of comet formation (Chapter 3) but the various mechanisms proposed generally involve comet formation on $\sim10^{4}$–10^{6} yr timescales. For comets condensing over longer timescales, a more secure source of radioactive heating would appear to come from the nuclides ^{232}Th, ^{238}U and ^{40}K.

Wallis (1980) considered a heat-conduction model with ^{26}Al. By integration of the diffusion equation he showed that a comet of radius ~3–6 km containing a plausible fraction of ^{26}Al would melt in the centre. A central low pressure water-droplet mixture was envisaged, insulated by a surrounding icy shell, the water phase being maintained against re-freezing for timescales of ~My. Yabushita (1993) explored models involving longer-lived radioactive elements such as ^{232}Th, ^{238}U (with half-life > 1 Gy). With appropriate mass fractions and on the assumption of a very low thermal diffusivity (10–100 times lower than for crystalline-ice) it was shown that relatively large comets, $r > 150$ km would be liquefied. Yabushita considered further the interesting possibility of chemistry leading to prebiotic molecules occurring within large liquid comets.

We now consider an approximate model of a radiogenically-heated comet to compute its thermal evolution. Merk and Prialnik (2003) derive the equation

$$C_p \frac{dT}{dt} = \tau^{-1} X_0 H e^{(-t_0+t)/\tau} - \frac{3\kappa T}{R^2 \rho} \tag{7.5}$$

where T is the average temperature, C_p is the average specific heat, τ is the half-life of the radioactive element, H is the radioactive heat input per unit mass, κ is the average thermal conductivity, X_o is the mass fraction of ^{26}Al and R is the cometary radius. Here t_o (assumed to be ~1 My) is the time between injection of ^{26}Al and the formation of comets, and t is the time after the comet was formed.

We assume $C_p = 10^3$ J/kg, $\rho = 0.3$ g/cm^3, $d = \kappa/(\rho C_p) = 10^{-6}$ m^2/s, for crystalline ice, and the previously assumed values of X_o and H relating to ^{26}Al. The adoption of values close to those appropriate to crystalline ice can be justified because the transition from amorphous ice to crystalline ice occurs at temperatures of ~120 K (Merk and Prialnik, 2003). Any transport of energy through convection would be stalled once percolating water vapour re-condensed and effectively sealed pores in the cometary body.

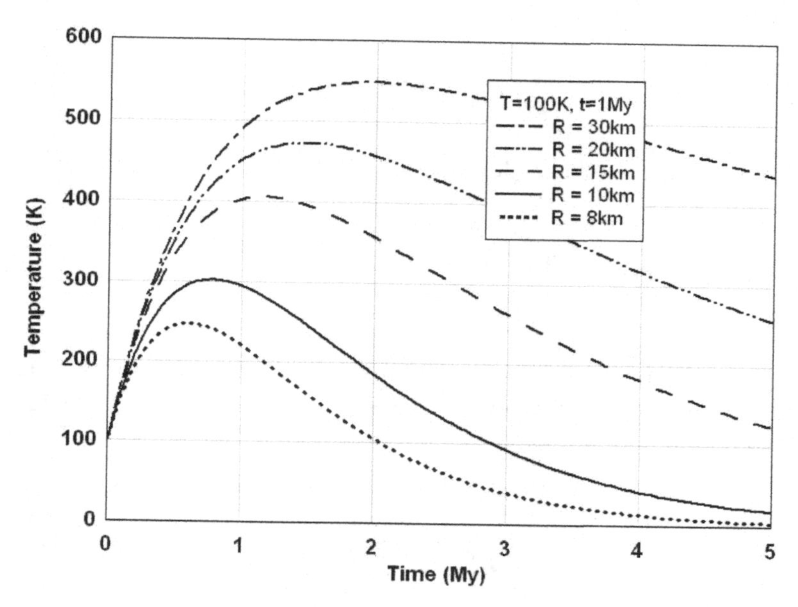

Fig. 7.1 Calculation of the average internal temperature of a comet accumulated 1 My after incorporation of ^{26}Al as a function of the subsequent time. Temperatures >273 K are artifacts of the idealised model. The temperature would flatten and subsequently decline after 273 K is reached.

Figure 7.1 shows the computed temperature by numerically solving Eq. (7.5) with T_o = 100 K. For comets with radii $R > 10$ km the temperatures reach 273 K at times $t = t_1 < 0.5$ My. The subsequent thermal evolution $T > 273$ K, $t > t_1$ as calculated according to Eq. (7.5) is of course unrealistic. Once the temperature T_m = 273 K is reached further heat released from radioactive decay is absorbed by the melting ice and T does not continue to rise. The energy balance equation at later times is given by

$$H_m \frac{dY}{dt} = X_1 H\tau^{-1} e^{-t/\tau} - \frac{3\kappa T_m}{\rho R^2} \tag{7.6}$$

where Y refers to the volume fraction of the comet melted at time t, and X_1 is the remaining fraction of ^{26}Al when the melting temperature is first reached given by

$$X_1 = X_0 \exp\left(-t_1/\tau\right) \tag{7.7}$$

The above equations due to Jewitt *et al.* (2007) lead to the solution

$$Y(t) = H_m^{-1}\left[X_1 H\left(1 - e^{-t/\tau}\right) - \frac{3\kappa T_m}{\rho R^2}\, t\right] \tag{7.8}$$

which is easily evaluated for specified values of the parameters. Figure 7.2 shows the volume fractions calculated for the two cases R = 12 km and 15 km.

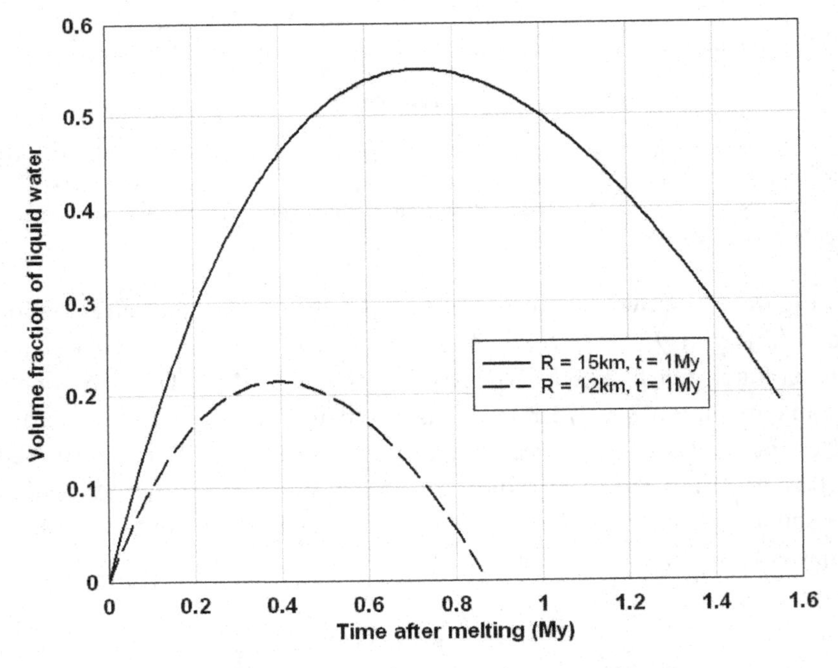

Fig. 7.2 Volume fractions of water sustainable after melting first occurs.

It is seen here that for smaller comets, even when they have radiogenically produced liquid cores, volume fractions and durations are less than for larger comets. However, for a plausible set of parameters we see that primordial water in comets is possible even for comets as small as 12 km. These results are generally in agreement with the trends shown in calculations by other authors (Jewitt *et al.*, 2007). However, uncertainties of the precise limiting comet size for liquid formation persist and are attributable to uncertainties in relevant thermal constants and/or detailed physical processes involved in the transport of energy through the comet.

Once a liquid core is established heat sources from exothermic recombination of radicals and organic chemicals would produce more melting (Hoyle and Wickramasinghe, 1985). Heat production will also follow from the hydration of rock particles. For a serpentine producing reaction

$$Mg_2 SiO_4 + MgSiO_3 + 2H_2 O \leftrightarrow Mg_3 Si_2 O_5(OH)_4$$

the energy released per unit mass is comparable to the latent heat of fusion of ice, 3.3×10^5 J/kg, thus permitting runaway melting near the boundary of an ice layer (Grimm and McSween, 1989; Jakosky and Shock, 1998).

The importance of radiogenic heating for panspermia is that a minute fraction of viable microbes that might be included in the initial cometary material would replicate exponentially on timescales very much shorter than the timescale for refreezing. The mixtures of water, minerals and organics in warm liquid cometary cores provide an ideal culture medium for anaerobic microorganisms.

7.3 Evidence of Present-day Melting

Aqueous alteration of minerals in C1 carbonaceous chondrites inferred from geochemical and textural analyses provides compelling evidence for liquid water in their parent bodies (McSween, 1979). If comets are accepted as being parent bodies of a large class of carbonaceous chondrites (Section 4.4.1 and Hoover, 2005) liquid water in comets needs no further proof. Cometary dust (interplanetary dust particles) collected in the atmosphere has also exhibited aqueous alteration in hydrated states. They include components of clays, serpentines and carbonates sharing many mineralogic similarities with chrondritic material (Brownlee, 1978).

The first space probes to a comet in 1986 explored comet Halley at close range and made discoveries that were at odds with the hard-frozen icy conglomerate model. As predicted by Hoyle and Wickramasinghe (1986) the comet's surface turned out to be exceedingly black, with albedo ~3%, blacker than soot and hardly visible. Mass spectroscopy with *Giotto* instruments confirmed a chondritic composition of comet dust.

Infrared maps of comet Halley by *Vega 1* showed temperatures in the range $T \approx$ 320–400 K at 0.8 AU, considerably higher than the sublimation temperature of water-ice in space ~200 K. More recently-studied comets have led to similar conclusions on the whole. Comet 19/P Borrelly observed by *Deep Space 1* in 2001 at 1.4 AU showed a very dark surface with an average albedo ~3%, but with local spots even darker with albedos of less than 0.008 — darker than carbon black. Again surface temperatures were found to be in the range 300–340 K, with no traces of water. The surface temperatures 300–340 K at 1.4 AU are appropriate to solar heating with little thermal conduction. Comet 81/P Wild 2 observed by *Stardust* also showed similar low albedos, as did comet Tempel 1. Indeed, numerous ground-based observations of active comet nuclei in the ~10–20 μm waveband confirm that albedos generally lie in the range 0.02–0.04.

The observation of comets showing the presence of insulating crusts that resist high sub-solar temperatures prompted the model that is now described. Our objective will be to discover whether liquid water can be produced by solar heating at some depth below the surface crust.

We first consider a non-rotating stationary comet under sub-solar insolation with a view to solving the heat conduction equation

$$\frac{\partial T}{\partial t} = \kappa \nabla^2 T \quad (7.9)$$

which in a one-dimensional infinite plane approximation simplifies to

$$\frac{\partial T}{\partial t} = \kappa \frac{\partial^2 T}{\partial z^2} \quad (7.10)$$

We envisage a layered one-dimensional structure to emerge over a depth of ~1 m of comet material due to outgassing near perihelion passage. Over this depth porous organic lag material would accumulate following percolation/sublimation of volatiles, but the outermost 1–2 cm would inevitably comprise a sun-burnt insulating crust that can frequently crack and reseal under the pressure ~10 mb of escaping water vapour and volatiles. Organic volatiles reaching the surface would continually re-seal and strengthen this 'mobile' outer crust.

Consider a three-layer structure with burnt insulating material at the top from $z = 0$ to $z = d_1$ with thermal conductivity $\kappa_2 = 0.17$ W/(mK) (corresponding to bitumen), a layer with $d_1 < z < d_2$ possessing conductivity $\kappa_1 = 1.7$ W/(mK) (organic-ice mix). The possibility of liquid water is considered for a depth $z > d_2$.

The energy balance is controlled by the receipt of solar radiation at r AU

$$F_0 = \frac{1.4}{(r/AU)^2} \quad kW/m^2 \quad (7.11)$$

on the comet's surface, and the thermal re-emission

$$\sigma \varepsilon_{IR} T^4 \quad (7.12)$$

which yields

$$\sigma \varepsilon_{IR} T^4 = \frac{(1-A) F_0}{(r/AU)^2} - \Phi_{down} \tag{7.13}$$

Here ε_{IR} refers to the mean infrared emissivity averaged over the thermal re-emission waveband and

$$\Phi_{down} = \kappa \frac{dT}{dz} \tag{7.14}$$

is the net downward flux of radiation during the day and A is the albedo, taken to be 0.03.

If water is mixing and convecting in a steady state of heat flow the solution satisfies

$$\Phi_{down} = \kappa_1 \left(T_0 - T_i\right)/d_1 = \kappa_2 \left(T_i - 273\right) \tag{7.15}$$

where T_0 is the maximum outer temperature of the comet, T_i is the temperature between the outermost burnt layer and the organic crust, and 273 K is the interface temperature between the organic crust and the water layer.

The outer temperature T_0 is calculated from Eqs. (7.13) and (7.15), and a schematic temperature resulting from such a calculation is set out in Fig. 7.3.

We now consider a hypothetical layered structure as described above, but with the outer surface exposed to diurnal variations as it would be for a tumbling comet. In the case of comet Halley the tumbling period is 90 hr. For a comet rotating with period τ, the temperature at a depth x below an area of thin insulating crust is approximated by a harmonic function. We use an infinite half-plane model to solve the heat conduction equation

$$\frac{\partial T}{\partial x} = \alpha \frac{\partial^2 T}{\partial x^2} \tag{7.16}$$

where $\alpha = \kappa / C\rho$, κ being the thermal conductivity, C the thermal capacity and ρ the density. With values appropriate to the organic comet material $C = 0.8$ J/cm^{3}K, $\kappa = 1.7$ W/mK we have $\alpha = 0.003$ m^2/h for use in Eq. (7.9) and a thermal skin depth ~30 cm.

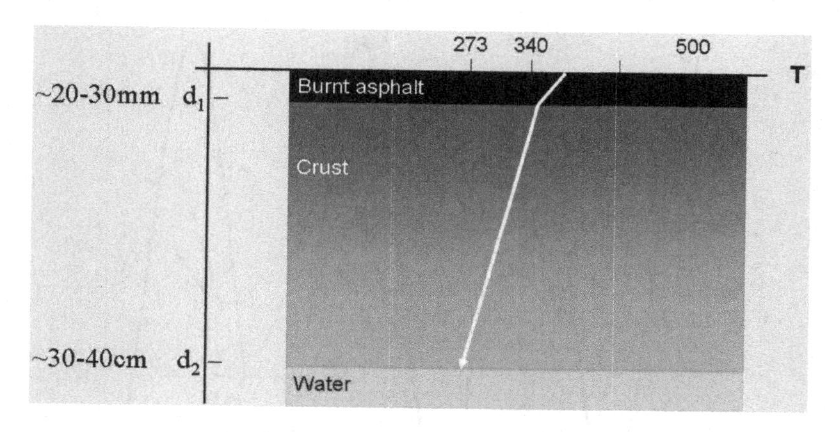

Fig. 7.3 A schematic temperature profile of a comet.

For an infinite half-plane maintained at the surface $x = 0$ with an oscillatory (sine-wave) temperature variation of amplitude T_{ampl} and period τ about a mean temperature T_{mean}, (determined by long-term solar insolation) the analytical solution of Eq. (7.9) is

$$T(x, t) = T_{mean} + T_{ampl} \cos\left(\sqrt{\frac{\pi}{\alpha\tau}} - 2\pi\frac{t}{\tau}\right) e^{-\sqrt{\frac{\pi}{\alpha\tau}}\,x}$$

$$+ \frac{2T_{initial}}{\sqrt{\pi}} \int_0^{x/2\sqrt{\alpha t}} e^{-\xi^2}\, d\xi \qquad (7.17)$$

where $T_{initial}$ is the constant initial temperature of the medium (Carslaw, 1921).

We solve Eq. (7.17) numerically with $\alpha = 0.003$ m^2/h as an appropriate value for the organic matrix underneath the burnt crust. Figures 7.4 and 7.5 show the variations of temperature with time at various depths below the insulating crust. Figure 7.4 is for a comet at 0.7

AU and Fig. 7.5 is one at 1.5 AU, where the maximum subsolar temperatures are determined by midday solar heating assuming an emissivity $\varepsilon = 1$.

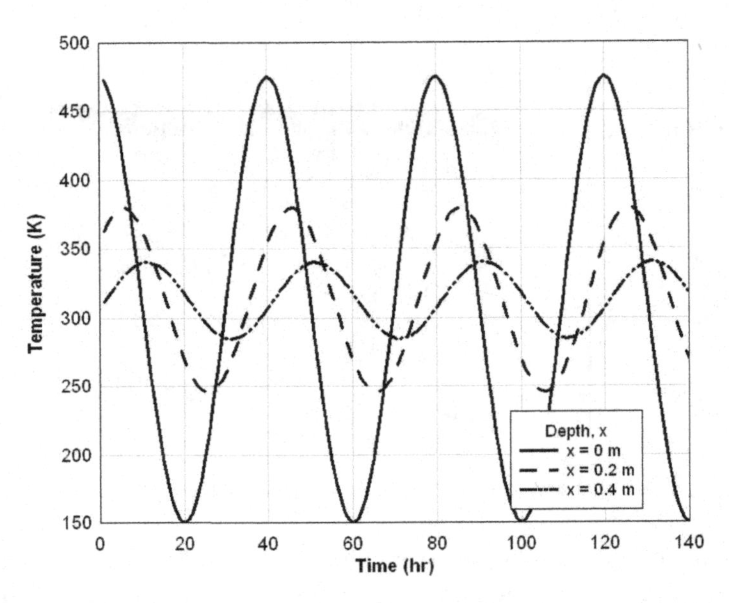

Fig. 7.4 Variations of temperature with time, at depth x beneath an insulating crust of comet with emissivity $\varepsilon = 1$ at 0.7 AU rotating with period 40 hr.

In Figure 7.4 we note that liquid water can persist below 0.2 m (e.g. comet Halley); in Fig. 7.5 (e.g. comet Tempel 1) no liquid water is possible. According to these calculations the generation of transient subsurface liquid water domains requires the temperatures at some depth to remain above a notional melting temperature ($T = 273$ K assumed, but could be lower, even 250 K for organic-water-ice mixtures) throughout a diurnal cycle. For perihelion distances <1 AU this could be feasible at depths ~0.3–0.5 m, subsurface pools persisting for days or weeks near perihelion. This provides ample time for a vast amplification of a population of dormant bacteria that may have been present (*cf.* Eq. (7.4), with typical doubling times of hours. The liquid pools themselves would be maintained by the tensile strength of the overlying layer which would be at least 100 times the water vapour pressure ~6 mb. If, however, gas pressure builds up due to metabolic activity the tensile strength would be insufficient to maintain equilibrium and the crust will be ruptured, leading to the release of gas and dust (Wickramasinghe, Hoyle and Lloyd, 1996).

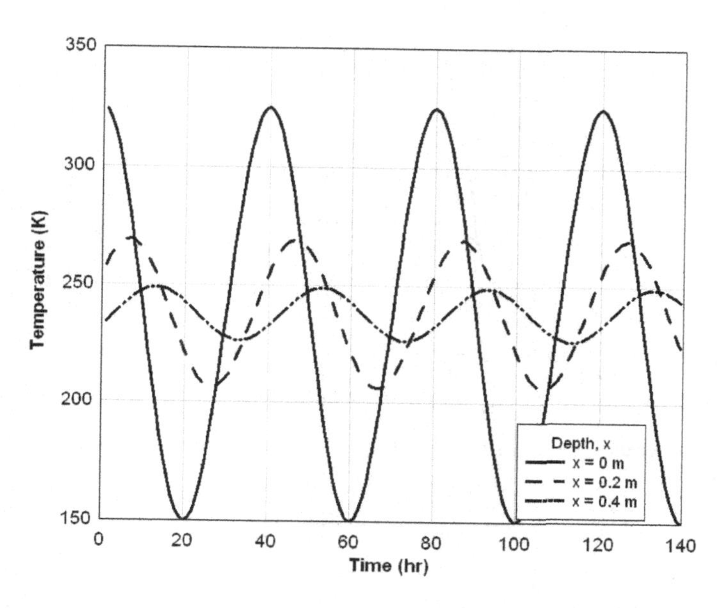

Fig. 7.5 Variations of temperature with time, at depth x beneath an insulating crust of a comet with emissivity $\varepsilon = 1$ at 1.5 AU rotating with period 40 hr.

7.4 Results from *Deep Impact*

The composition and structure of the crust and subcrustal layers of a comet came sharply into focus on 4th July 2005. NASA's *Deep Impact* mission dispatched a 370 kg projectile at a speed of ~10 km/s to crash head-on into comet Tempel 1. Large quantities of gas and grains were expelled to form an extended plume and coma. Spectra of the coma in the near IR waveband 10 minutes before and 4 minutes after the impact obtained by A'Hearn *et al.* (2005) showed a sharp increase of emission by CH bonds in organic dust over the waveband 3.3–3.5 μm relative to an approximately constant water signal (Fig. 7.6 a,b).

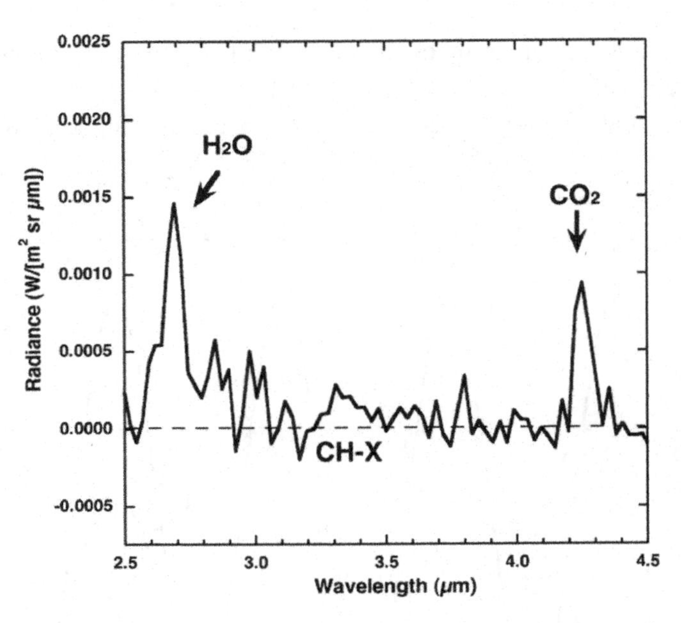

Fig. 7.6a Spectrum of coma of comet Tempel 1 ten minutes before impact (A'Hearn *et al.*, 2005).

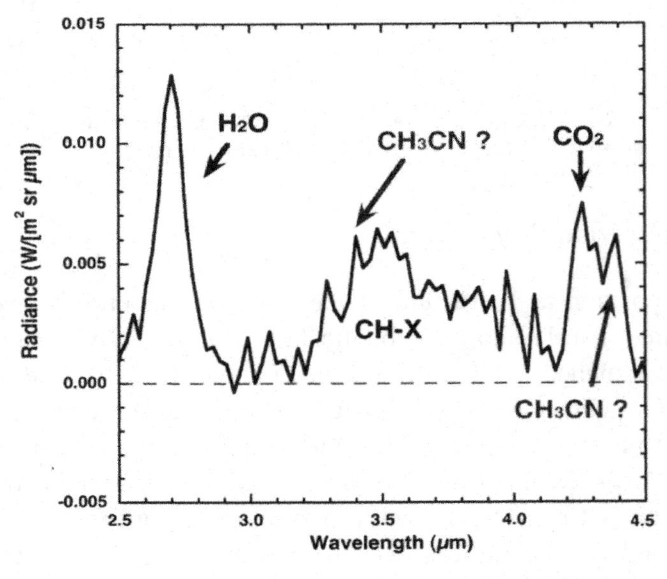

Fig. 7.6b Spectrum of coma of comet Tempel 1 four minutes after impact (A'Hearn *et al.*, 2005).

The excess radiation over this waveband in the post-impact plume that cannot be modelled by inorganic coma gases are best explained on the basis of degraded biologic-type organic material. We considered the type of organic dust discussed by Hoyle and Wickramasinghe for explaining the mean 3.3–3.5 μm emission profile for comet Halley and other comets (organic/biologic and asphalt-like material presumed to result from its degradation). Figure 7.7 shows the normalised emissivity calculated for such a model compared with the observations of the post-impact plume of Tempel 1 on 4th July 2005.

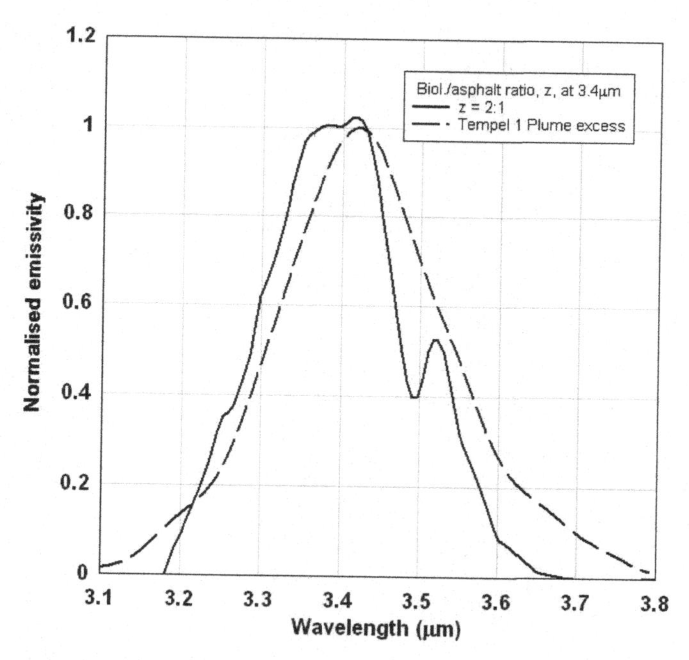

Fig. 7.7 Normalised emissivity near 3.4 μm for a mix of 'normal' cometary dust matching biological material with asphalt (solid line). For comparison is plotted the 'unexplained' component of emission from the plume following impact on 4th July 2005 (dashed line).

We see that, with a 2:1 contribution of biological material to asphalt in the emissivity at 3.4 μm, an approximate match with the data is obtained. This is satisfactory considering the fact that asphalt is likely to be an over-simplified representation of degraded biomaterial.

Observations of Tempel 1 some 20 hours following impact using the *Spitzer Telescope* yielded 5–37 μm spectra generally similar to the spectra of comet Hale-Bopp taken by ISO (Infrared Space Observatory). The Hale-Bopp data was interpreted by Wickramasinghe and Hoyle (1999) on the basis of a model involving olivine and biological material in the mass ratio 1:10. To model the Tempel 1 data we considered mixtures of olivine and biological material with normalised absorbance curves as given in Fig. 7.8. This figure also shows a normalised absorbance curve from a standard 'laboratory' clay as given by Frost, Ruan and Kloprogge (2000).

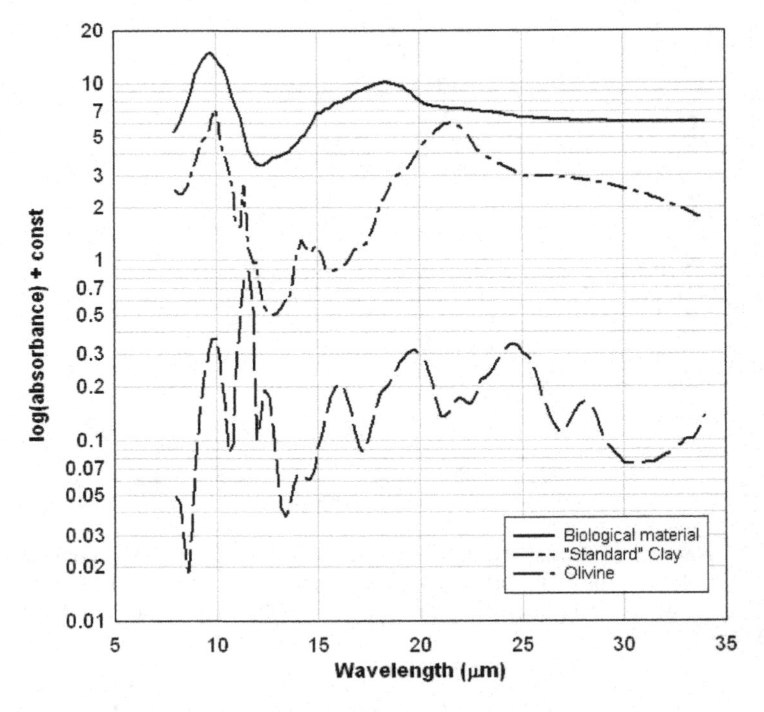

Fig. 7.8 Absorbance data used for the calculations.

The absorbance data for olivine and biomaterial are taken from compilations by Hoyle and Wickramasinghe (1999). The two materials were conceptually mixed with opacity contributions at their peak maxima in the ratio organic: olivine = z:1. Since for olivine/clay the maximum mass absorption coefficient near the 9 μm peak is over 5 times that of the less well-ordered organics, the corresponding mass ratios are near $5z$:1.

In Fig. 7.9 we plot the normalised flux curves for emission from biological/olivine mixtures for grain temperatures T = 350 K with z = 10 and 3. The dashed curve is the normalised flux obtained from Tempel 1 (Lisse *et al.*, 2006). For z = 3 corresponding to an organic to mineral mass ratio in excess of 10:1, the overall agreement is seen to be satisfactory. Temperatures much below 300 K appear to be ruled out on account of the observed slope of the flux curve from 5–33 μm.

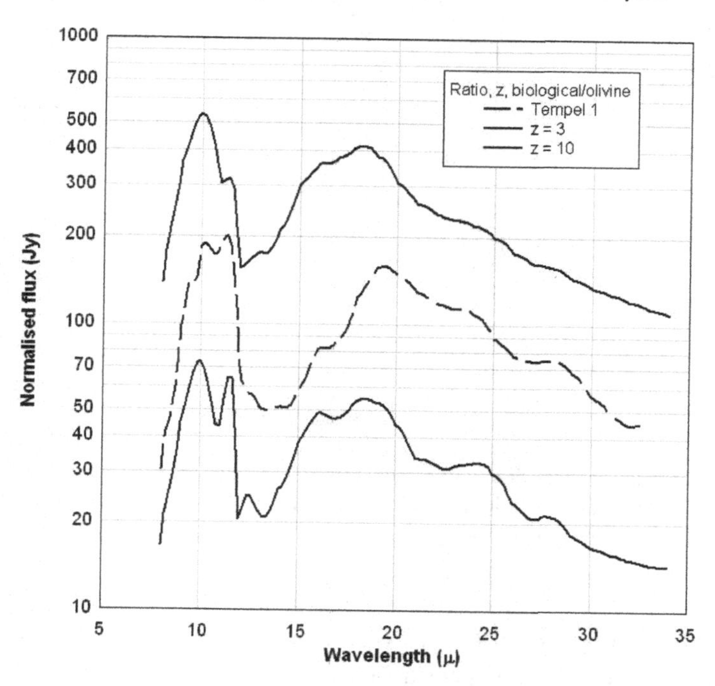

Fig. 7.9 Normalised flux due to mixtures of biological material and olivine. z is the ratio of opacity contributions (biological: olivine) near the 9 μm peaks. The dashed curve represents data for Tempel 1.

Figure 7.10 displays computed flux curves for mixtures of clay and biological grains, again heated to 350 K. Very similar relative abundance conclusions can be drawn in this instance, with mass ratios of biological to clay again well in excess of 10:1. The temperature 300–350 K appears to be necessary to match the underlying continuum of the Tempel 1 dust spectrum.

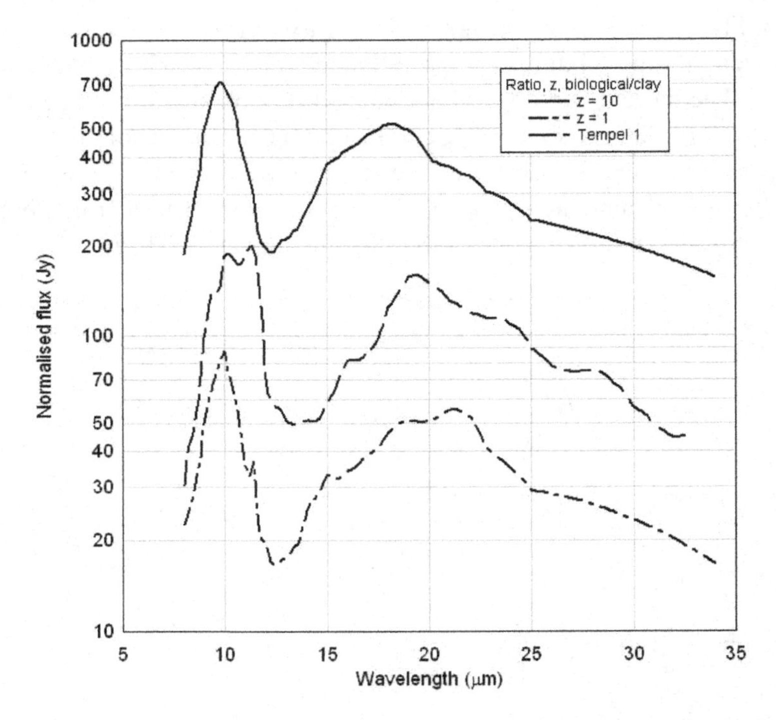

Fig. 7.10 Normalised flux due to mixtures of biological material and clay. z is the ratio of opacity contributions (biological: clay) near the 9 μm peaks. The dashed curve represents data for Tempel 1.

Figure 7.10 shows the situation for T = 250 K, indicating that the relative heights of the 10 and 20 μm peaks are inconsistent with the observational data. The spectra of clays tend to be somewhat variable one to another reflecting their idiosyncratic mineral configurations. However, fine structure in the 8–11 μm absorption band centred near 9,

10 and 11 μm can in general be regarded as signifying a clay, and these indeed have been found in the higher resolution spectra of Tempel 1 following the *Deep Impact* event.

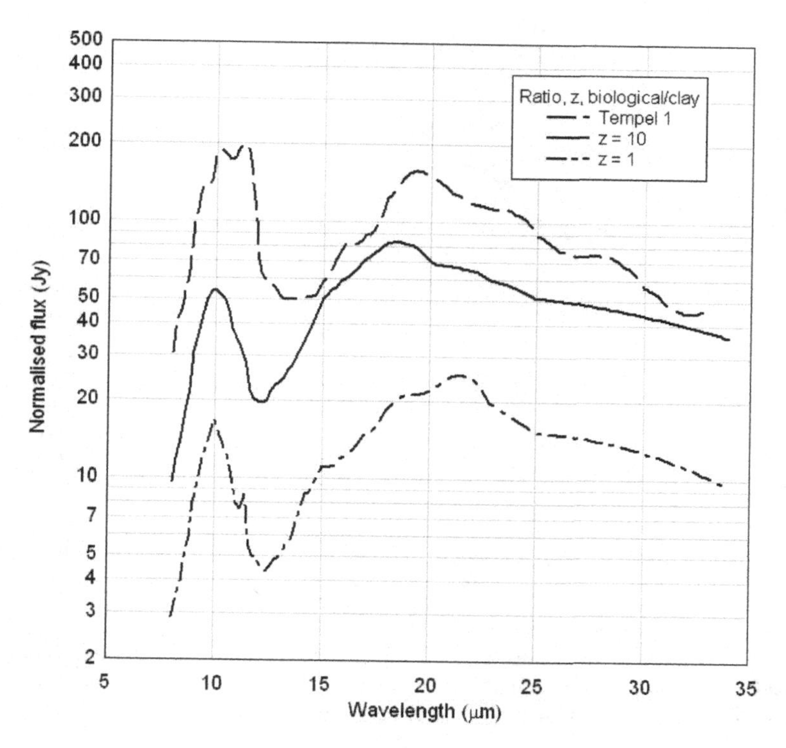

Fig. 7.11 Normalised flux due to mixtures of biological material and 'standard clay'. z is the ratio of opacity contributions (biological: clay) near the 9 μm peaks. The dashed curve shows data for Tempel 1.

Over the 8–13 μm wavelength interval we combined the standard clay spectrum given by Frost and Kloprogge with three other clay spectra, yielding the normalised emissivity plotted in Fig. 7.12. The average emissivity spectrum (post-impact) of Tempel 1, shows good agreement with clay at 9 and 10 μm, but a significant excess centred near 11.2 μm (Lisse *et al.*, 2006). This 'missing' flux has been attributed to

PAHs although individual compact PAHs tend to possess a 11.2 μm feature that is too narrow to fit the bill.

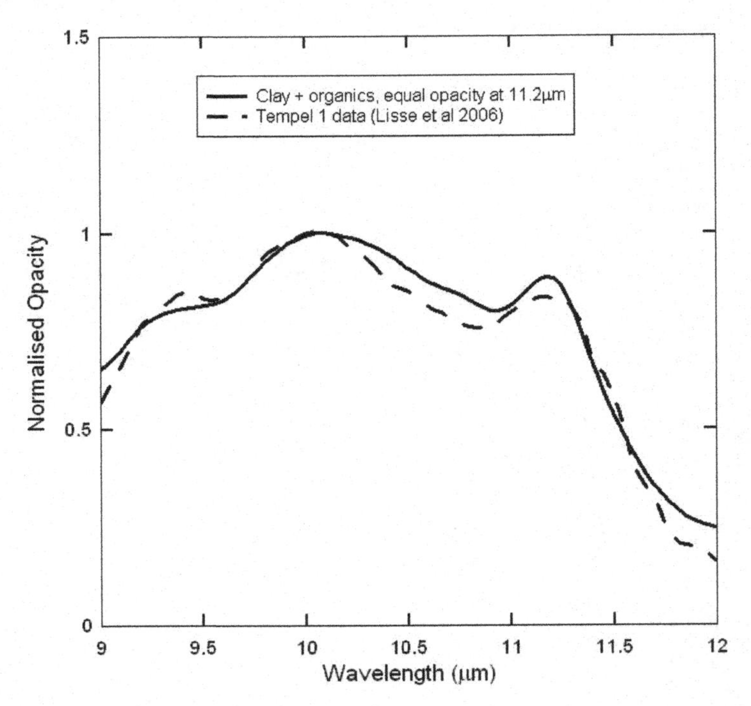

Fig. 7.12 Normalised opacities near 10 μm due to mixtures of clays and biological aromatic molecules. Comparable contributions at 11.2 μm from clays and aromatic molecules fit most of the long-wave opacity requirements.

Hoyle and Wickramasinghe (1991) had argued that an ensemble of biological aromatic molecules might explain the UIBs (unidentified infrared bands) in interstellar spectra, and their set of 115 biological aromatics do indeed provide an 11.2 μm band with a half width at maximum intensity of ~0.3 μm. Such as absorption can fit a Lorentzian profile such as calculated in Fig. 7.13. Emission in such a 11.2 μm band contributing an equal flux to that from clay at the band centre adds up to the curve shown in Fig. 7.12, exhibiting satisfactory overall agreement with the observations.

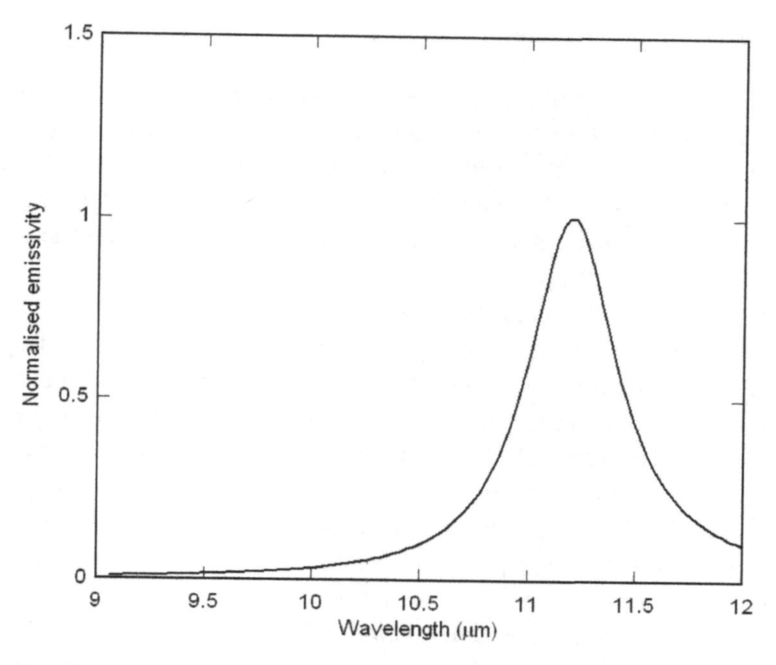

Fig. 7.13 Normalised opacity attributed to 115 biological aromatic molecules.

We thus confirm the conclusions already reached by Lisse *et al.* (2006) that clay minerals, complex organics and PAH-type material are present in the post-impact ejecta from Tempel 1. Earlier claims in the last decade of similar materials in IDPs (interstellar dust particles) and in C1 carbonaceous chondrites cannot now be ignored. Furthermore, since clay minerals are thought to form only in contact with liquid water, water in comets now appears a virtual certainty. Although terrestrial clay is formed by erosional weathering of rocks, with the possible intervention of microbial activity, in astronomical situations other processes could operate. Interstellar dust that is incorporated in comets would have included a component of submicron silica particles that were produced in the mass flows from cool stars. Such particles are likely to be amorphous at the outset, but when they come in contact with liquid water in the melted phases of comets, hydration leading to clay structures could result.

7.5 Frozen Lake Surfaces

Transient sub-surface pools with organic nutrients and viable microbes imply that continuing biological activity could occur in comets. The pools in our model are maintained a few tens of centimetres below a sealed outer crust that is continually regenerated due to condensation of diffusing organic materials. Such a crust can easily withstand the 6 mb pressure needed to support a water interface, but sporadic rupturing of areas of crust could occur leading to episodes of outgassing. Such rupturing could occur due to impacts on the surface, or by the build-up of pressure due to resumption of biological activity in sub-surface niches as proposed by Wickramasinghe *et al.* (1998). High resolution images of Tempel 1 show smooth terrains that could be understood as refrozen lakes, where impacts dislodged overlying loose crustal material (Fig. 7.14). The fraction of surface where H_2O ice has been detected is, however, less than 3%, possibly connected with exposed, refrozen subsurface lakes. Two circular plugs of ice surrounded by 'moats' could be impact melts that have developed stable attached crusts, while the original large craters around them have fully eroded. The smallest craters seen on Tempel 1 have central hills that are likely to be consolidated stable crusts left by impact melts as the surrounding material erodes more quickly.

Comets are not the only small bodies in the solar system that show evidence of frozen water surfaces. The main belt asteroid Ceres, which has a mean diameter of 952 km, eccentricity $e = 0.08$ and semimajor axis 2.77 AU, has recently revealed a water-ice surface as indicated in the spectrum shown in Fig. 7.15 (Vernazza *et al.*, 2005). It is argued that the water-ice surface has been recently emplaced because the sublimation life-time at 2.8 AU is relatively short. Here then is another instance of a recently-exposed subsurface lake as in a comet.

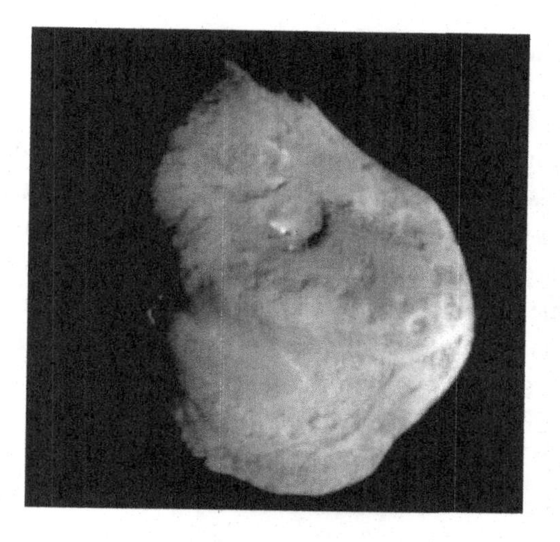

Fig. 7.14 High resolution image of Tempel 1 (Courtesy: NASA).

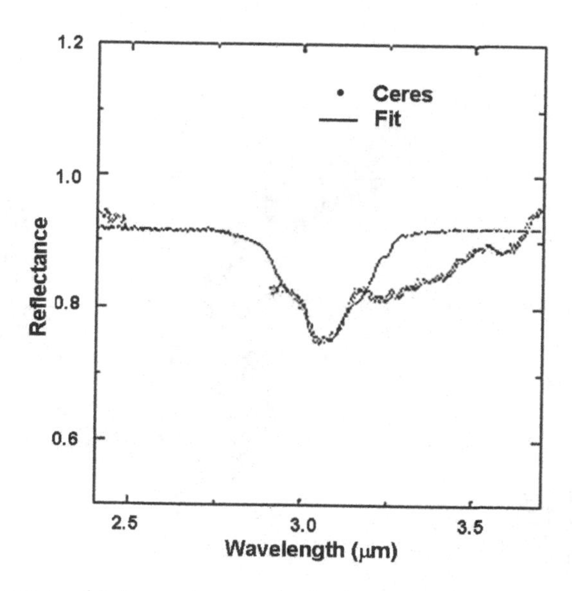

Fig. 7.15 Spectrum of 1 Ceres showing the 3.06 μm absorption band due to crystalline water ice. The thin line is a model, where the data at longer wavelengths require another absorber (from Vernazza *et al.*, 2005).

7.6 Microbial Fossils in Carbonaceous Meteorites

The possible existence of microbial fossils in carbonaceous meteorites has had a long history. Original claims by Claus and Nagy (1963) of discovering such microfossils (organised elements) were mostly discredited on the grounds of possible contamination. Later investigations undertaken by H.D. Pflug using ultrathin sections of the Murchison meteorite revealed a similar profusion of organised elements, which became difficult to rule out convincingly as contaminants. Figure 7.16 shows a structure within the meteorite compared with an SEM of a recent microorganism *Pedomicrobium*, obtained by Hans Pflug (see ref. in Hoyle and Wickramasinghe, 1982). Microprobe spectral analyses carried out by Pflug confirmed that the Murchison structures contain molecules that are consistent with biodegraded material.

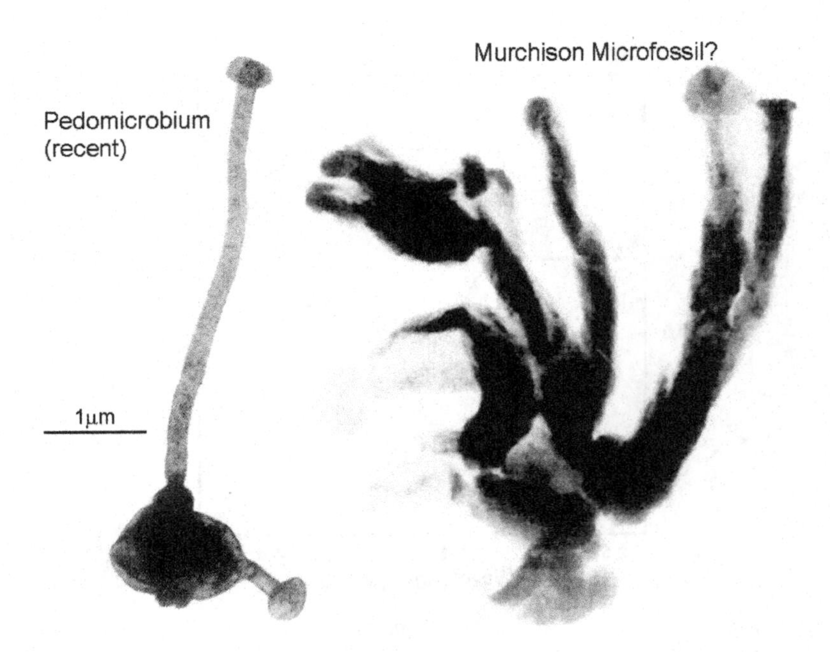

Fig. 7.16 Carbonaceous structure in the Murchison meteorite isolated by Pflug compared with a recent microorganism.

Even the careful work of Pflug failed to carry conviction, and more recently an extensive programme of similar studies has been undertaken by R.B. Hoover and his collaborators at NASA (Hoover, 2005). Hoover has documented images of entire colonies of fossilised microbes securely lodged within the mineral matrix of the meteorite, as for instance in Fig. 7.17. Again microprobe studies and elemental analyses further confirm a biological provenance for these structures.

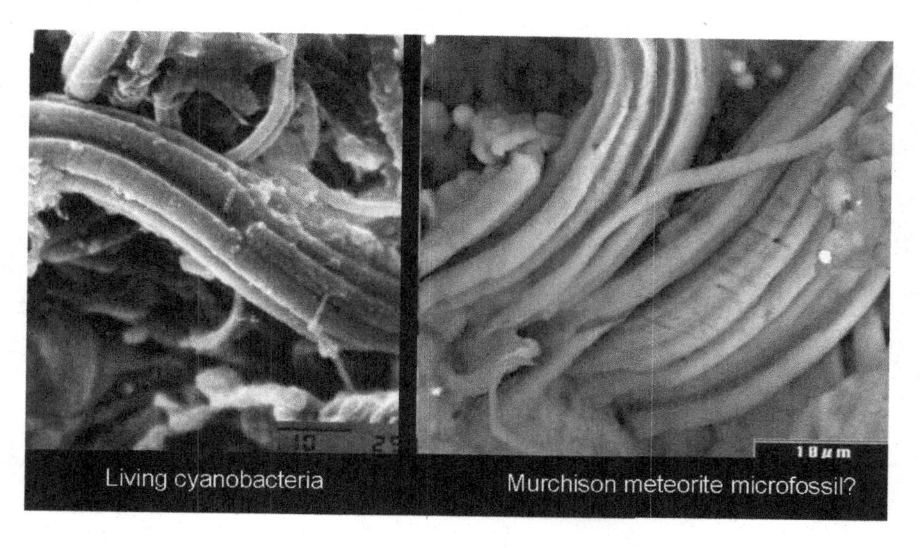

Fig. 7.17 A structure in the Murchison meteorite (Hoover, 2005) compared with living cyanobacteria.

Carbonaceous meteorites could be associated with many types of parent bodies, but comets would appear to head the list of candidates. The dynamical evolution of cometary orbits with transient solar heating could provide the conditions needed for the resumption of microbial activity at the surface and thus explain the occurrence of microbial fossils in carbonaceous meteorites.

Chapter 8

Origin of Life

8.1 Preamble

The problem of the origin of life and the origin of the Universe are the most difficult unsolved problems in science and are likely to be inextricably linked. Although in the latter instance a 'concordance' theory exists for a 'big-bang' origin of the Universe, alternative contenders are still very much in the field, and controversies continue to rage (Hoyle *et al.*, 2000; Arp *et al.*, 1990). Although the connection of these two problems is rarely discussed, a link there must surely be. For instance, if the Universe had no beginning the possibility that life was an ever-present entity must surely be considered. The recent discovery of infrared signatures consistent with biological aromatic molecules in external galaxies, extending to a gravitationally-lensed galaxy at a redshift of $z = 0.83$ supports the view that life could be widespread over a large fraction of the volume of the observable Universe (Wickramasinghe *et al.*, 2005). Moreover, galaxies of even higher redshifts (e.g. in the Hubble telescope deep-field images) exhibit morphologies very similar to present-day galaxies, as well as high levels of metallicity and chemical evolution attesting to their suitability at least for harbouring life.

Figure 8.1 shows the nebula IC1396 in the constellation of Cepheus which includes many molecular clouds and regions of active star formation. The conspicuous dark globules seen here represent dust clouds undergoing gravitational collapse, the first stage in the formation of stars.

The solar system formed from a fragment of interstellar cloud similar to the dark globules seen in Fig. 8.1. It collapsed under its own self-gravitation some 4.6 billion years ago (Sections 3.1 and 3.2), the collapse possibly being triggered by the explosion of a nearby supernova, which also had the effect of injecting radioactive nuclides such as ^{26}Al into the embryonic protoplanetary disc. The condensation of the planets is thought to have been achieved through a series of steps listed in Chapter 3.

Fig. 8.1 A star-forming region IC1396 in the constellation of Cepheus (Courtesy: Canada-Franco-Hawaii telescope — J.-C. Cuillandre).

As described therein models of a hierarchical accumulation and coalescence of planetesimals at different heliocentric distances indicate that the formation of the inner planets was completed in a few hundred thousand years, while the condensation of the outer planets took a few million years. A fraction of the icy planetesimals (cometary bodies) from which the outer planets formed contributed to the present-day Oort cloud of comets, and a fraction was gravitationally perturbed into the inner planetary system to collide with planets, including Earth.

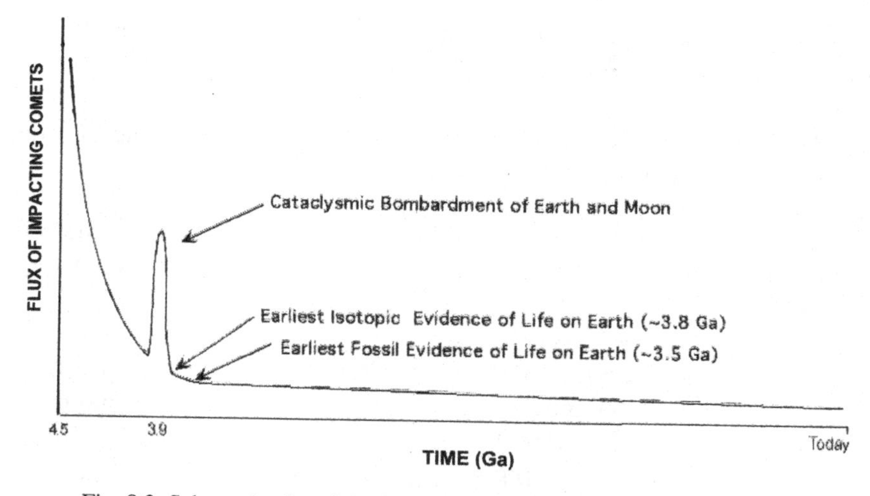

Fig. 8.2 Schematic plot of the frequency of cratering impacts on the Earth.

Figure 8.2 shows a schematic diagram of the impact rates of comets on both the Earth and the Moon, showing a declining average flux with an episode of enhanced collisions at ~4.4 Gy BP, defining the so-called Hadean epoch, possibly coinciding with a moon-forming collision event. It is generally believed that the comets impacting Earth during the Hadean epoch contributed to a large fraction of the planet's volatiles, including water.

The primordial soup theory of the origin of life was developed independently by the Russian biochemist A.I. Oparin and the English biologist J.B.S. Haldane (Oparin, 1953). Here the Earth's early atmosphere is assumed to consist of a mixture of H_2O, CH_4, NH_3 with an overall reducing character. These gases are then assumed to dissociate into radicals by the action of solar UV and lightning, and in their recombination a trickle of organics rains down into the oceans to form a primordial soup. Chemical reactions in the 'soup' are then postulated to lead to the formation of a simple self-replicating living system through a sequence of steps. A problem with this theory emerged as soon as it was discovered that the Earth's primordial atmosphere was of an oxidising — not reducing — nature, in which case no organics can be formed.

Since comets are known to contain organic molecules relevant to life two options remain for the origin of life on the Earth:

(1) The primordial soup theory operates on the Earth after comets dump organic molecules into the lakes and oceans.

(2) A similar theory operates within the set of radioactively heated comets, with chemistry leading to life occurring in liquid cometary cores.

Estimates of the odds *against* life emerging from non-life anywhere are highly uncertain, but must surely range from being superastronomical to hugely astronomical (Hoyle and Wickramasinghe, 1980). The recognition that life has an information content too vast to be derived from random processes has led to many discussions of incremental evolutionary steps, starting from simpler self-replicating systems and ultimately leading to protein–DNA-based life. The currently popular RNA–world hypothesis is based on the evidence that short stretches of RNA molecules act in the capacity of enzymes, catalysing their continuing replication (Woese, 1968; Gilbert, 1986). The clay model (Cairns-Smith, 1966; Cairns-Smith and Hartman, 1986) uses the repeating lattice structures of clay particles and their catalytic properties of converting simple organic molecules in aqueous solution into complex biopolymers. The lattice structure of clay serves as the first template in an undefined progression to life. RNA molecules are known to persist and replicate upon clay surfaces, thus combining aspects of the RNA–world theory with the clay model (Cha *et al.*, 2000).

In view of the high abundance of silicon in the Galaxy the clay world model might well have a special role to play in a cosmic context. The transition from any of these intermediate systems to the final DNA–protein-based cellular life form is still in the realm of speculation.

The difficulty of finding any evidence of the relics of prebiology in the geological record has been an handicap for all Earth-based theories of the origin of life. The suite of organics present in interstellar clouds and in comets would seem to direct our search for origins away from Earth to more and more distant parts of the Universe. At the very least the organic molecules needed for life's origins are much more likely to have been generated in a cosmic context rather than being formed *in situ* on Earth. Moreover, it is now becoming clear that life arose on Earth almost at the very first moment that it could have survived. During the period from about 4.3–3.8 Gy ago (the Hadean Epoch) the Earth suffered an episode of heavy bombardment by comets and asteroids. Rocks dating back to the tail end of this epoch reveal evidence of an excess of the lighter isotope ^{12}C compared with ^{13}C pointing to the action of microorganisms that preferentially took up the lighter isotope from the environment.

Harold Urey (1952) and Stanley Miller (1953) attempted to mimic the processes discussed by Haldane and Oparin: by sparking a mixture of H_2O, CH_4, NH_3 in the laboratory they obtained small yields of amino acids, nucleic acid bases and sugars. The success of the Urey–Miller experiment led to the expectation that it was only a matter of time before the next steps from biochemical monomers to life could be demonstrated in the laboratory. Despite over half a century of effort, this goal has proved stubbornly elusive.

If one accepts the calculations showing grotesquely small *a priori* probabilities for the transition of non-life to life (Crick and Orgel, 1973; Hoyle and Wickramasinghe, 1980, 1982), it would appear that only two options remain open:

1. The origin of life on Earth was such an extremely improbable event that if it happened here, it will effectively not be reproduced elsewhere.

2. Or, a very much bigger system and a longer timescale were involved in an initial origination event after which life was somehow transferred to Earth.

We ourselves prefer the latter option, whilst others may choose to disregard the probability argument as insecure, and assert that life must of necessity arise readily by an undiscovered set of processes whenever and wherever the right conditions are found. Spontaneous generation is revived!

8.2 Cometary Interiors as Incubators of Early Life

In the context of recent studies of comets, the clay theory of the origin of life merits serious consideration. The *Deep Impact* mission to comet Tempel 1 (4 July 2005) revealed spectroscopic evidence for a mixture of organics and clay particles that were carried away in the post-impact plume (Lisse *et al.*, 2006; A'Hearn *et al.*, 2005). The presence of a range of complex hydrocarbon molecules that might be considered appropriate monomers for prebiotic chemistry were also indicated in infrared spectra — similar to molecules found in material recovered from comet Wild 2 in the *Stardust* mission (Sandford *et al.*, 2006). Thus the required aqueous environment for this model, containing high concentrations of organics and clay particles, is now established to occur in comets.

It is curious that modern biology displays a residual affinity to silica and highly ordered silica structures. Biogenic silica often displays ring structures and ultra-microcrystalline order. Diatoms use silica in the form of complex siliceous polymers overlain with protein templates in their cell walls (Mopper *et al.*, 1973). The formation of solid silica structures with precisely controlled morphologies appear to be directed by DNA coded proteins (Cha *et al.*, 2000), thus supporting the idea that a reversal of logic could lead from highly ordered silica morphologies (a primitive genome) to a DNA based informational system. In higher life as well a relic memory of a silica connection could be seen to persist, leading in some instances to human diseases such as silicosis (Grant *et al.*, 1992).

The liquid water condition in cometary interiors may be maintained by radiogenic heat sources such as ^{26}Al over timescales of $\sim 10^6$ yr as we have seen in Chapter 7. For large comets (say >100 km in diameter) the liquid phase may be an order of magnitude longer. The surfaces of clay particles, which are estimated to have a total volume of 10^3 km^3 in a typical comet may thus provide an ideal platform for prebiotic chemical evolution to commence.

The emergence of chirality — a preference for one optical isomer over another in sugars and amino acids — is another factor that favours a cometary origin of life. Cataldo (2007) has pointed out that life cannot be conceived without chirality, and that the prolonged action of cosmic rays on prebiotic molecules in the interstellar medium, or the action of sunlight in a terrestrial context, would act against the preservation of chirality of asymmetric molecules. However, in cometary interiors, the only photons present are γ-ray photons arising from the decay of radiogenic nuclides. Cataldo (*loc. cit.*) finds that, in the presence of the expected radioactive decay in cometary interiors, a significant fraction of chiral molecules would survive over several Gy. In one specific case that was studied, the optical activity was enhanced by the gamma radiation, yielding a polymer which could act as a template for prebiotic chemistry. Thus cometary interiors seem to favour the chirality selection found in present-day biology.

The recent discovery of a diverse microbial ecosystem at a depth of 2.8 km in the Archean crust, evidently sustained by the energy of radioactive decays over millions of years, gives additional credence to the idea that life could originate under similar conditions within comets (Lin *et al.*, 2006). The *a priori* case for a clay-based model for an origin of life in a clay-organic-water comet somewhere in the Galaxy, as against an origin in a diminutive setting on the Earth, is manifestly strong.

8.3 Comparison with a Terrestrial Origin of Life

On the present-day Earth aqueous suspensions of clay particles in conjunction with organics persist mostly in hydrothermal vents, the total volume of which could hardly be in excess of 10^3 km^3. Clay present elsewhere in the crust, formed through weathering of rocks and accumulating in shallow pools, would have an erratic short-term persistence, drying up seasonally, or with colloidal particles sedimenting under gravity on shorter timescales. Soon after the end of the Late Heavy Bombardment, the deposition rate of clay in the Earth's crust would have been minimal, as is indicated by the thin layers of clay found at the top of the oldest Pre-Cambrian sediments (Ziegler and Longstaffer, 2000). With organic molecules supplied erratically by comet impacts, their concentrations in the oceans will necessarily be exceedingly low — too dilute presumably for any prebiotic chemistry to proceed (Hazen,

2005). Transiently high concentrations will of course be achieved by evaporation near the margins of lakes and seas. But these high concentrations would also have a short-term seasonal persistence. All these factors would constitute a severe limitation for an Earth-based theory of the origin of life.

In Chapter 7 we saw that there is evidence for clay particles in comets, and these were obviously not formed by rock erosion processes. We pointed out that interstellar silicate grains included within the melted cores of comets could undergo aqueous alteration to become clay-like structures.

At the presumed time of life's origin on Earth we consider an effective depth of 2 m of clay covering 1% of the Earth to be an optimistic upper limit for the total clay volume ($\sim 10^4$ km^3) that is transiently available for catalytic chemistry. Taking $\sim 10^{12}$ comets with mean radius 10 km in an initial Oort cloud, and if a comet contains 30% by volume of clay, the total volume of clay particles in solar system comets is 10^{15} km^3. This gives a factor 10^{10} in favour of comets, on the basis of relative clay volume alone. Whereas the average persistence of shallow clay pools and hydrothermal vent concentrations of clay can range from 1 yr to ~ 100 yr, a cometary interior provides a stable, aqueous, organic-rich environment for $\sim 10^6$ yr. Thus there is another factor of at least $10^6 / 100 = 10^4$ in favour of solar system comets, raising their relative chances to 10^{13}. But given that panspermia takes place on an interstellar scale we have to multiply this number by the number of solar system clones in the Galaxy. If 10% of G-dwarf stars are endowed with planetary systems and Oort clouds of comets, the final number for the Galaxy is 10^{24}. Thus the mass and stability of suitable cometary environments overwhelms any which may have existed on the early Earth: if life was first assembled in a clay system, the odds against the clay being terrestrial are $\sim 10^{24}$ to 1 against. Similar considerations apply to other proposed prebiotic pathways, such as those of the PAH (Hazen, 2005), lipid (Szostak *et al.*, 2001) or peptide (Carny and Gazit, 2005) worlds.

Liquid water, organic molecules and surfaces on which catalytic reactions can take place are likely prerequisites for the emergence of life. All these requirements are met within comets and there is no compelling reason why cometary prebiotic molecules need to be brought to Earth before they can be assembled. Cometary interiors provide a much better option if the totality of comets in the Galaxy is taken into account (Napier *et al.*, 2007). Mechanisms discussed in the literature (such as the clay world of Cairns-Smith) work far better in liquid cometary interiors than they do in the harsh environment of the early Earth. And in terms of their total mass and surface area for catalytic reactions, stability of environment and high nutrient concentration, comets are overwhelmingly favoured.

Chapter 9

Expanding Horizons of Life

The history of ideas about humanity's place in the universe consists of a long series of retreats from a geocentric position. From this perspective the argument that life is here, therefore it must have started here, can be seen as the last bastion of geocentrism; it is in any case surely flawed.

The link between comets and life has in fact a long history: comets were at once revered and feared in ancient cultures, often carrying the belief that they bring pestilence and death. If comets do indeed harbour microorganisms and also collide with the Earth from time to time to bring death and destruction, these ancient beliefs may have more than a grain of truth to commend them. This might be unpalatable to those who dismiss all ancient ideas as arising from ignorance and superstition.

The major objection to panspermia on anything other than a local scale has always been the apparent impossibility of transporting living or dormant microorganisms between the stars. As we have seen, however, this problem has receded as our knowledge of extremophiles and of the solar system and Galactic environments has advanced. Transmission of self-replicating organisms into star-forming regions is feasible; only a handful need to land on a receptive planet for life to take hold; and to judge by Earth history, life seems difficult to dislodge once it has started.

The panspermia concept does not directly address the problem of the origin of life. It does, however, greatly extend the scope of putative originating environments and removes us from what may be the bottleneck of geocentric theories. In fact, some of the most promising ideas about the origin of life on Earth require conditions which are more easily met in the interiors of large comets. Relative to terrestrial settings, such stable, aqueous, nutrient-rich environments are overwhelmingly predominant in the Galaxy or even within our solar system.

The first obvious pointers to our cosmic ancestry come from astronomical arguments that cannot be refuted. Following the work of Burbidge, Burbidge, Fowler and Hoyle (1957) on nucleosynthesis we know that all the atoms of which life is made — C, O, N, P, S, Na, K etc. — were synthesised in nuclear reactions in stars, and expelled into the interstellar medium in supernova explosions. Along with the atomic species ejected into space in this way, some refractory minerals are also formed in stellar environments and are ejected into interstellar clouds. Chemistry taking place either in the interstellar gas (ion–molecule reactions) or on the surfaces of refractory dust grains can lead to the formation of interstellar molecules including organic molecules, but all such processes tend to be self limiting. To proceed from simple organic molecules formed in this way to biochemical polymers and life in the interstellar medium appears to be exceedingly unlikely. On the other hand, relatively simple organic molecules delivered into protoplanetary nebulae — sites where comets and planets may form — could undergo further transformations in aqueous environments on the surfaces, or in the interiors of planetary bodies, leading to the first self-replicating living system. Whilst prebiotic chemical evolution can be easily envisaged with evolution towards the most reproducible, stable and radiation resistant molecular aggregates, the progression to life involves superastronomically large probability obstacles.

A 'continuity principle' adopted by biologists states that there is 'no unbridgeable gap' between inorganic and living matter; each stage in evolution develops continuously from the previous one, at each stage there is a continuous path backwards to the prebiotic state and forward to modern organisms. This principle, stated in various ways by many biologists, is purely conjectural and has no overwhelming logic to commend it.

As we have argued in this book the best setting for a transition from prebiotic chemistry to chemical evolution leading to life probably requires an aqueous environment with high concentrations of organics in contact with clay-like particles (Cairns-Smith, 1966). While this combination of conditions has been discussed at length by proponents of the clay-RNA theory of the origin of life in a purely terrestrial context, we have argued that even a single comet offers a better prospect (Napier, Wickramasinghe and Wickramasinghe, 2007). On the Earth adequate concentrations of organics in conjunction with clay particles can occur only near the margins of lakes and oceans, where seasonal evaporation could lead to enhanced concentrations of organics. Not only is the total volume of such a concentrated terrestrial 'soup' limited, its duration is seasonal and moreover, the periods of time when clay particles remain in suspension, stirred by winds and currents and without sedimenting under gravity are even shorter. We argued that evidence for clay particles exists at least in some comets (e.g. comet Tempel 1) and the radioactively heated interiors of these comets provide a far better option for the first origin of life. The chirality of biomolecules has a natural explanation in terms of their origin in the ^{26}Al-heated interiors of large comets but is problematic otherwise.

Whereever this highly improbable transformation took place it will be spread across galactic and intergalactic distances with relative ease. The absolute survival limits for bacteria and genetic material under cryogenic interstellar conditions are still under vigorous debate (Cowan and Grady, 2000; Horneck *et al.*, 2002). All the indications are that an adequate viable fraction for panspermia can be maintained everywhere, possibly removing the requirement for multiple origins at least within a single galaxy.

We argued in this book that comets not only harbour primitive microbial life, but they amplify and distribute life on a galactic scale. Major disruptions of a life-bearing Oort-type cloud of comets take place due to the combined effects of the vertical Galactic tide and close encounters with molecular clouds. Such encounters lead to an increase in the flux of comets driven into the inner planetary system, comets that can collide with 'inhabited' planets like the Earth. The cratering record on the Earth bears testimony to such impacts occurring at intervals ranging from 20–40 My. A fraction of the unshocked material excavated from impact craters and ground up into dust can be expelled from the solar system due to the action of radiation pressure. During episodes of molecular dust encounters viable bacterial spores released directly from comets can also be expelled by radiation pressure to reach speeds of ~30 km/s. Such bacterial dust would reach nascent planetary systems in the approaching molecular cloud on timescales as short as 1 My, timescales over which viability of microbial life is well assured.

The next step in our logic is to suppose that extrasolar planetary systems are no different from our own solar system in being endowed with Oort-type clouds of comets, and that their major disruption episodes also occur on timescales of ~20–40 My. Alien planetary systems may or may not have planets on which Earth-type life can evolve, and so an unbroken chain of transmission of planet-borne life might not be assured. However, the redistribution of primitive microbial life directly via comets must surely take place. Every planetary system like our own would seed embryonic planetary systems and Oort-type clouds within 'passing' molecular clouds. If one planetary system seeds $1 + \varepsilon$ others with an effective time-step of 40 My, the number of planetary systems seeded in the age T of the Galaxy is

$$\sim (1 + \varepsilon)^{T/40\,\mathrm{My}}$$

With $T \approx 10^{10}$ yr for the time during which life became first established in the galaxy, and with $\varepsilon \approx 0.1$, we obtain a total number of 2×10^{10} seeded planetary systems. This calculation simply illustrates the inevitability of the spread of microbial life within a galaxy, and how a galaxy could be 'colonised' from an initial source in a relatively short time.

While life could spread throughout the habitable zone of the Galaxy via the processes discussed in this book, could it be transmitted between galaxies? If so, the hunt for life's origins may extend to the Local Group, the Local Supercluster or even the observable Universe. Connecting 'lanes' of stellar and interstellar matter between galaxies are not uncommon (see Fig. 9.1) with, for example, the Magellanic Stream joining us to the Large Magellanic Cloud at ~300,000 light years distance.

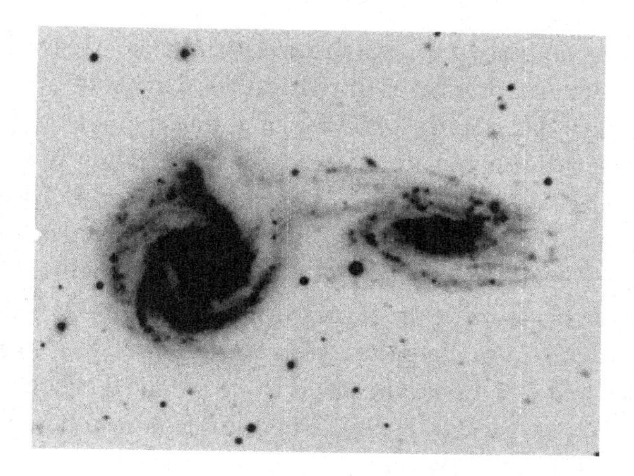

Fig. 9.1 Spiral galaxies NGC5427 and NGC5426 connected by a bridge of gas and stars. This photo (negative image) was taken in 1966 by H. Arp using the 200-inch Palomar telescope. Since then many such bridges between galaxies have been discovered (Courtesy: Harlton Arp).

Spreading microbial life within a cluster of galaxies can presumably take place with relative ease. As pointed out by Hoyle and one of the present authors (Hoyle and Wickramasinghe, 2000) bacterial clumps attached to iron whiskers could be expelled entirely from galaxies. When metallic vapours are cooled in the laboratory, condensation occurs into threads or whiskers with typical diameters of 0.01 μm and lengths about a millimetre, giving a very large ratio of length to diameter. Such metallic particles are extremely strongly repelled by radiation in the far infrared region of the spectrum, and since molecular clouds in galaxies emit radiation strongly in the far infrared, whiskers can be repelled from galaxies into extragalactic space at speeds of upwards of a few thousand km/s — say 3×10^{8} cm s^{-1}.

Galaxies over a radius ~30 Mpc could thus be reached from a galaxy where life originated in less than 10^{10} years. One might typically expect about a million galaxies to be thus reached, and the products of biological evolution in a million galaxies could be mixed in this way. The expansion of the Universe could do the rest — the expansion essentially doubling the cosmological region through which life is now distributed every 10^{10} years (The scale length in the universe expanding with time t as $\exp(H_0t) = \exp(t/13.6 \text{ Gy})$). The first stars, on the conventional picture, originated at redshift $z \sim 15$, about 50 My after the Big Bang. These had masses ~100 $M_\odot$, creating the heavy elements and dispersing them through supernova explosions. The first galaxies of Milky Way size appeared at 400 My, at redshift $z \sim 10$. If life had originated in the universe at say redshift $z = 0.8$ where we saw evidence of biological aromatic molecules in Chapter 2, a single origination event is all that is needed to bring the legacy of life to present-day galaxies.

The idea that life predated the solar system is given strong support from the fact that life first appears in the geological record at the very first moment it can survive during a period of intense bombardment by comets. It was most likely then that comets not only delivered organic molecules onto the Earth and water to form the oceans, they brought fully fledged life carrying an evolutionary history that predated even the formation of the Galaxy itself. If so we would expect to find similar life taking root on other planetary bodies of our solar system.

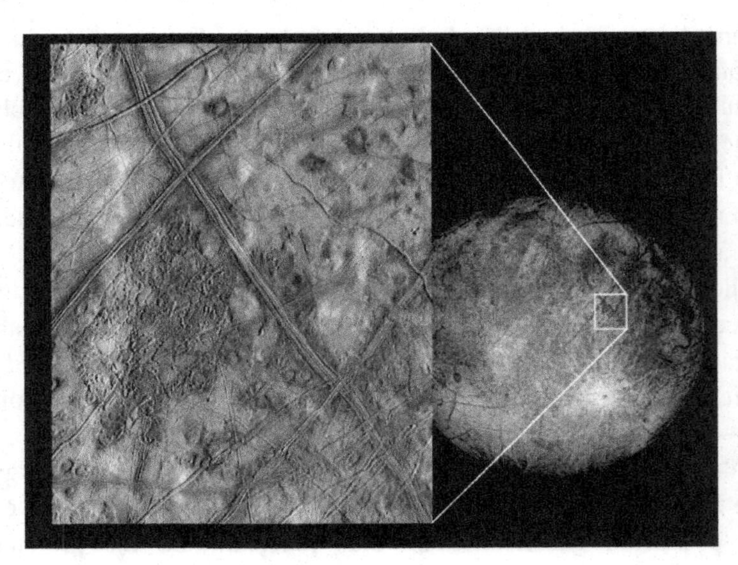

Fig. 9.2 Multicracked surface of Europa with orange pigments having spectral characteristics of bacterial pigments (Courtesy: NASA).

Whilst it is common to look for signs of prebiology on satellites of the outer planets, e.g. Titan, the exciting possibility would be to look for evidence of life. Similarly, on the Jovian satellite Europa, there is tantalising evidence of microorganisms inhabiting the tidally heated sub-surface oceans that have been inferred to exist — the orange 'pigments' demarcating the cracks (see Fig. 9.2) showing spectroscopic similarities to bacterial pigments. And Venus, although written off for life because of its high surface temperature, could harbour microbial ecologies 50 km above the surface in the clouds where temperatures are 70 K and pressures are ~1 bar (Wickramasinghe and Wickramasinghe, 2008).

The case for contemporary life on Mars is also not being ruled out by any means and, as Gil Levin points out, the *Viking* landers in 1976 may well have discovered life that is only now coming to be recognised (Levin, 2007). And the *Mars Express* discovery of atmospheric methane is most plausibly explained as a biogenic product. All these matters are still being vigorously debated, and no doubt would be resolved in the not too distant future. In our view the impediment to present progress is the insistence on *de novo* origins of life and pre-biotic evolution in every location where life is looked for.

The cometary panspermia model developed in this book presents several challenges to conventional biological thought. Darwinian evolution, survival of the fittest, would proceed not in a closed system, but in an open system permitting, as well as depending upon, periodic injections of new genetic material from space. If the same genotypes are periodically re-introduced to the planet from a frozen and non-evolving reservoir within comets, conventional views about phylogenetics may need to be revised. Horizontal gene transfers that are now widely discussed in a purely terrestrial context must logically be extended to a much bigger galactic system. The idea that the phylogenetic tree provides an unerring time sequence of biological evolution then becomes seriously flawed.

In summary the arguments presented in this book support the view that comets are responsible not only for the arrival of life on Earth (and on similar exosolar planets) but for its subsequent evolution as well. Comets are the carriers, amplifiers and distributors of life. Our cosmic ancestry is firmly linked to comets.

Bibliography

A'Hearn, M.F., Belton, M.J.S., Delamere, W.A. *et al.*, 2005. *Science*, **310**, 258.

Abraham, M. and Becker, R., 1950. *The Classical Theory of Electricity and Magnetism.* 2nd ed. London: Blackie.

Adams, F.C. and Spergel, D.N., 2005. *Astrobiology*, **5**, 497.

Allen, C.W. 2000. *Astrophysical Quantities,* New York: Springer.

Allen, D.A. and Wickramasinghe, D.T., 1981. *Nature,* **294**, 239.

Al-Mufti, S., 1994. *PhD thesis*. University College, Cardiff.

Arp, H.C., Burbidge, G., Hoyle, F., Narlikar, J.V. and Wickramasinghe, N.C., 1990. *Nature*, **346**, 807.

Arrhenius, S., 1903. *Die Umschau,* **7**, 481.

Arrhenius, S., 1908. *Worlds in the Making*. London: Harper.

Bahcall, J.N., 1984. *Astrophys. J.*, **276**, 169.

Bailey, M.E., 1990. *In:* D. Lynden-Bell and G. Gilmore, eds. *Baryonic Dark Matter.* Dordrecht: Kluwer, 7.

Bailey, M. E. and Emel'yanenko, V. V., 1998. *In:* M.M.Grady *et al.*, eds. *Meteorites: Flux with Time and Impact Effects.* London: Geol. Soc. London, 11.

Bailey, M. E. and Stagg, C.R., 1988. *MNRAS*, **235**, 1.

Bailey, M.E., Clube, S.V.M. and Napier, W.M., 1990. *The Origin of Comets.* Oxford: Pergamon.

Barranco, H.J., 2009. *ApJ*, in press.

Becquerel, P., 1924. *Bull. Soc. Astron.*, **38**, 393.

Beichman, C.A., Bryden, G., Gautier, T.N. *et al.*, 2005. *Astrophys. J.*, **622**, 1160.

Bidle, K., Lee, S-H., Marchant, D.R. and Falkowski, P.G., 2007. *Proc. Nat. Acad. Sci. USA,* **104**(33), 13455.

Biryukov, E.E., 2007. *Solar System Research*, **41**, 211.

Biver, N., Bockelée-Morvan, D., Crovisier, J. *et al.*, 2002. *Earth, Moon & Planets*, **90**, 323.

Bohler, C., Nielson, P.E. and Orgel, L.E., 1995. *Nature,* **376**, 578.

Böhren, C.F. and Wickramasinghe, N.C., 1977. *Astrophys. Sp. Sci.*, **50,** 461.

Böhnhardt, H., Mumma, M.J., Villanueva, G.L. *et al.*, 2008. *ApJ*, **683**, L71.

Bonnell, I.A. and Bate, M.R., 2006. *MNRAS*, **370**, 488.

Bradley, J.P., Brownlee, D.E. and Fraundorf, P., 1984. *Science*, **223**, 56.

Brasser, R., Duncan, M.J. and Levison, H.F., 2008. *Icarus*, **196**, 274.

Bridges, J.C., Changela, H.G., Carpenter, J.D. and Franchi, I.A., 2008. *Proc. 39th Lunar and Planetary Science Conference*, 10–14 March 2008 League City Texas, 2193.

Brownlee, D.E., 1978. *In:* T. Gehrels, ed. *Protostars and Planets.* Tuscon: Univ. of Arizona Press, 134.

Burbidge, E.M., Burbidge, G.R., Fowler, W.A., and Hoyle, F., 1957. *Rev. Mod. Phys.*, **29**, 547.

Burch, C.W., 1967. *Symp. Soc. Gen. Microbiol.*, **17**, 345.

Burchell, M.J., Mann J.R. and Bunch, A.W., 2004. *MNRAS*, **352,** 1273.

Butler, R.P., Wright, J.T., Marcy, G.W. *et al.*, 2006. *ApJ*, **646**(1), 505.

Byl J., 1986. *Earth, Moon, and Planets*, **36**, 263.

Cairns-Smith, A.G., 1966. *J. Theor. Biol.*, **10**, 53.

Cairns-Smith, A.G. and Hartman, H., eds., 1986. *Clay Minerals and the Origin of Life.* Cambridge University Press.

Campins, H. and Swindle, T.D., 1998. *Meteoritics & Planetary Science,* **33**, 1201.

Cano, R.J. and Borucki, M., 1995. *Science*, **268**, 1060.

Carny, O. and Gazit, E., 2005. *FASEB Journal*, **19**, 1051.

Carslaw, H.S., 1921. *The Conduction of Heat.* London: Macmillan.

Cataldo, F., 2007. *Int. J. Astrobiol.*, **6**, 1.

Cataldo, F. and Keheyan, Y., 2003. *Int. J. Astrobiol.*, **2**(1), 41.

Cataldo, F., Keheyan, Y. and Heymann, D., 2004. *Origins of Life and Evol. of Biospheres,* **34,** 13.

Cataldo, F., Keheyan, Y. and Heymann, D. 2002. *Int. J. Astrobiol.*, **1**, 79.

Cha, J. N., Stucky, G.D., Morse D.E. *et al.*, 2000. *Nature*, **403**, 289.

Chambers, J. E. and Migliorini, F., 1997. *Bulletin of the American Astronomical Society*, **29**, 1024.

Christensen, E.A., 1964. Acta Path. et Microbiol. Scandinavia, **61**, 483.

Claus, G., Nagy, B. and Europa, D.L., 1963. *Ann. NY Acad. Sci.,* **108**, 580.

Cleaves, H.J. and Chalmers, J.H., 2004. *Astrobiology,* **4**(1), 1.

Clemett, S.J., Maechling, C.R., Zare R.N. *et al.*, 1993. *Science*, **262**, 721.

Clube, S.V.M. and Napier, W.M., 1984. *MNRAS*, **208**, 575.

Clube, S.V.M. and Napier, W.M., 1996. *QJRAS*, **37**, 617.

Coulson, S.G., 2004. *Int. J. Astrobiol.*, **3**(2), 151.

Coulson, S.G., 2009. *Int. J. Astrobiol.*, **8**(1), 9.

Coulson, S.G. and Wickramasinghe, N.C., 2003. *MNRAS,* **343**, 1130

Cowan, D and Grady, M., 2000. *Microbiol. Today*, **27**, 174.

Crick, F.H.C. and Orgel, L.E., 1973. *Icarus,* **19**, 341.

Crovisier, J., Leech, K., Bockelée-Morvan, D. *et al.*, 1997. *Science*, **275**, 1904.

Darbon, S., Perrin, J.-M. and Sivan, J.-P, 1998. *Astron. and Astrophys.*, **333**, 264.

Darwin, C., 1859. On the Origin of Species by Means of Natural Selection. London: John Murray.

Davies, P.C.W., 2003. *Astrobiology*, **3**, 673.

Debye, P., 1909. *Ann. Physik*, **30**, 57.

Desvoivres, E., Klinger, J., Levasseur-Regourd, A.C. and Jones, G.H., 2000. *Icarus*, **144**, 172.

Diehl, R. and Timmes, F.X., 1997. *In*: C.D. Dermer *et al.*, eds. *AIP Conference Proceedings*, **410**, 218.

Draine, B.T., 2003. *Ann. Rev. Astron. Astrophys.*, **41**, 241.

Drobyshevski, E.M., 2008. *Icarus*, **197**, 203.

Drobyshevski, E.M., 1978. *Moon & Planets*, **23**, 339.

Edgeworth, K.E., 1949. *MNRAS*, **109**, 600.

Emel'yanenko, V. V. and Bailey M. E., 1998. *MNRAS*, **298**(1), 212.

Emel'yanenko, V.V., Asher, D.J. and Bailey, M.E., 2007. *MNRAS*, **381**, 779.

Fernández, J.A., 2005. Comets — Nature, Dynamics, Origin and their Cosmological Relevance. Dordrecht: Springer.

Fernández, J.A., 1997. *Icarus*, **129**, 106.

Fernández, J.A. and Ip, W.-H., 1987. *Icarus*, **71**, 46.

Fernández, J.A., Gallardo, T., Brunini, A., 2002. *Icarus*, **159**, 358.

Forrest, W. J., Houck, J.R. and Reed, R.A., 1976. *Astrophys. J.*, **208**, L133.

Forrest, W. J., Gillett, F.C. and Stein, W.A., 1975. *Astrophys. J.*, **192**, 351

Fouchard, M., 2004. *MNRAS*, **349**(1), 347.

Francis, P.J., 2005. *ApJ*, **635**, 1348.

Franck, S., Cuntz, M., von Bloh, W. and Bounama, C., 2003. *Int. J. Astrobiol.*, **2**(1), 35.

Frisch, P.C., 2007. *Space Sci. Rev.* **130**, 355.

Frost, R.L., Ruan, H. and Kloprogge, J.T., 2000. *The Internet Journal of Vibrational Spectroscopy*, **4**, 1.

Furton, D.G. and Witt, A.N., 1992. *Astrophys. J.*, **386**, 587.

Garcia-Sanchez, J., Weissman, P.R., Preston, R.A. *et al.*, 2001. *Astron. Astrophys.*, **379**, 634.

Gerola, H. and Schwartz, R.A., 1976. *ApJ*, **206**, 452.

Gezari, D.Y., Schmitz, M., Pitts, P.S. *et al.*, 1993. *Catalogue of Infrared Observations*. Washington: NASA.

Gies, D.R. and Heisel, J.W., 2005. *Astrophys. J.*, **626**, 844.

Gilbert, W., 1986. *Nature* **319**, 618.

Gladman, B., Michel, P. and Froeschle, C.H., 2000. *Icarus*, **146**, 176.

Gold, T., 1992. *Proc.Natl.Acad.Sci. USA*, **89**, 6045.

Grady, C.A., Woodgate, B., Bruhweiler, F.C. *et al.*, 1999. *Astrophys. J. Lett.*, **523**, L151.

Grant, D., Long, W.F. and Williamson, F.B., 1992. *Medical Hypotheses*, **38**, 46.

Greenblatt, C.L., Davis. A., Clement, B.G. *et al.*, 1999. *Microbial Ecology*, **38**, 58.

Gregory, P.H. and Monteith, J.L., eds., 1967. *Airborne Microbes – Symposium of the Society for General Microbiology*, Vol 17, Cambridge University Press.

Grimm, R.E. and McSween, H.Y., 1989. *Icarus*, **82**, 244.

Grün, E., Zook, H.A., Fechtig, H. *et al.*, 1985. *Icarus*, **62**, 244.

Guillois, O., Ledoux, G., Nenner, I. *et al.*, 1999. *Solid Interstellar Matter: The ISO Revolution.* Les Houches, No.11, EDP Sciences, Les Ulis.

Haldane, J.B.S., 1929. *The Origin of Life*. London: Chatto and Windys.

Hamilton, V.E., 2005. *Proc. 68th Annual Meeting of the Meteoritical Society*, 12-16 September 2005 Gatlinburg, Tennessee, 40, 5128.

Harris, M. J., Wickramasinghe, N.C., Lloyd, D. *et al.*, 2002. *Proc. SPIE*, **4495**,192.

Hazen, R.M., 2005. *Genesis*. Washington: Joseph Henry Press.

Heithausen, A., 2007. *In*: M. Haverkorn and W.M. Goss, eds. *Small Ionized and Neutral Structures in the Diffuse Interstellar Medium. ASP Conference Series*, **365**, 177. San Francisco: Astronomical Society of the Pacific.

Herbst, E. and van Dishoeck, E.F., 2009. *Ann. Rev. Astr. Astrophys.*, Sept, 2009 (in press).

Holmberg, J. and Flynn, C., 2004. *MNRAS*, **352**, 440.

Hoover, R. B., 2005, *In*: R.B. Hoover, A.Y. Rozanov and R.R. Paepe, eds. *Perspectives in Astrobiology. Amsterdam*: IOS Press, **366**, 43.

Hoover, R.B., Hoyle, F., Wickramasinghe, N.C. *et al.*, 1986. *Earth, Moon and Planets,* **35**, 19.

Horneck, G., 1993. Origins of Life and Evol. of Biospheres, **23**, 37.

Horneck, G. and Baumstark-Khan, C., eds., 2002. *Astrobiology. The quest for the conditions of life.* Berlin: Springer.

Horneck, G., Eschweiler, U., Reitz, G. *et al.*, 1995. *Adv. Space Res.,* **16**(8), 105.

Horneck, G., Mileikowsky, C., Melosh, H.J. *et al.*, 2002. *In*: G. Horneck, C. Baumstark-Khan, eds. *Astrobiology. The quest for the conditions of life.* Berlin: Springer.

Hoyle, F. and Wickramasinghe, N.C., 2000. *Astronomical Origins of Life: Steps towards Panspermia.* Kluwer Academic Press.

Hoyle, F. and Wickramasinghe, N.C., 1988. *Astrophys.Sp.Sci.*, **147**, 245.

Hoyle, F. and Wickramasinghe, N.C., 1989. *Astrophys.Sp.Sci.*, **154**, 143.

Hoyle, F. and Wickramasinghe, N.C., 1996. *Astrophys.Sp.Sci.*, **235**, 343.

Hoyle, F. and Wickramasinghe, N.C., 1979. *Diseases from Space,* London: J.M. Dent.

Hoyle, F. and Wickramasinghe, N.C., 1986. *Earth, Moon & Planets*, **36**, 289.

Hoyle, F. and Wickramasinghe, N.C., 1980. *Evolution from Space,* London: J.M. Dent.

Hoyle, F. and Wickramasinghe, N.C., 1981. *In:* C. Ponnamperuma, ed. *Comets and the Origin of Life*. Dordrecht: D. Reidel, 227.

Hoyle, F. and Wickramasinghe, N.C., 1978. *Lifecloud: the Origin of Life in the Galaxy.* London: J.M. Dent.

Hoyle, F. and Wickramasinghe, N.C., 1985. *Living Comets.* Cardiff: Univ. College Cardiff Press.

Hoyle, F. and Wickramasinghe, N.C., 1962. *MNRAS,* **124**, 417.

Hoyle, F. and Wickramasinghe, N.C., 1968. *Nature,* **217**, 415.

Hoyle, F. and Wickramasinghe, N.C., 1970. *Nature,* **226**, 62.

Hoyle, F. and Wickramasinghe, N.C., 1976. *Nature,* **264**, 45.

Hoyle, F. and Wickramasinghe, N.C., 1977. *Nature,* **270**, 323.

Hoyle, F. and Wickramasinghe, N.C., 1982. *Proofs that Life is Cosmic,* Mem. Inst. Fund. Studies Sri Lanka, No. 1 (www.panspermia.org/proofslifeiscosmic.pdf).

Hoyle, F. and Wickramasinghe, N.C., 1991. *The Theory of Cosmic Grains.* Dordrecht: Kluwer Academic Press.

Hoyle, F., Burbidge, G. and Narlikar, J.V., 2000. *A Different Approach to Cosmology.* Cambridge University Press.

Hoyle, F., Wickramasinghe, N.C., Al-Mufti, S., Olavesen, A.H., and Wickramasinghe, D.T., 1982. *Astrophys. Sp. Sci.,* **83**, 405.

Hoyle, F., Wickramasinghe, N.C. and Al-Mufti, S., 1982. *Astrophys. Sp. Sci.,* **86**, 63.

Hoyle, F., Wickramasinghe, N.C. and Pflug, H,D., 1985. *Astrophys. Sp. Sci.,* **113**, 20.

Hughes, D. W., 2001. *MNRAS,* **326**, 515.

Hughes, D. W., 2003. *MNRAS,* **338**, 999.

Hughes, D. W., 1996. *QJRAS,* **37**, 593.

Hulst, van de, H.C., 1946-49. *Rech. Astron. Obs. Utrecht,* XI, Parts I and II.

Hutsemékers, D., Manfroid, J., Jehin, E. *et al.,* 2005. *Astron. Astrophys.,* **440**, L21.

Ibadinov, Kh.I., 1989. *Adv. Spa. Res.,* **9**, 97.

Ibadinov, Kh.I., 1993. *In:* J. Stohl and I.P. Williams, eds. *Meteoroids and their Parent Bodies.* Bratislava: Astr. Inst. Slovak Acad. Sci., 373.

Ibadinov, Kh.I., Rahmonov, A.A. and Bjasso, A.Sh., 1991. *In:* R.L. Newburn, M. Neugebauer and J. Rahe, eds. *Comets in the Post-Halley Era.* Dordrecht: Kluwer, 299.

Igenbergs, E., Huedepohl, A., Uesugi, K.T. *et al.,* 1991. *IAF, 42nd International Astronautical Congress,* 5–11 October 1991 Montreal, 13.

Ishii, H.A., Bradley, J.P., Dai, Z.R. *et al.,* 2008. *Science,* **319**, 447.

Jakosky, B.M. and Shock, E.L., 1998. *J. Geophys.Res,* **103**, 19359.

Jayawardhana, R., Coffey, J., Scholz, A. *et al.,* 2006. *ApJ,* **648**, 1206.

Jeffers, S.V., Manley, S.P., Bailey, M.E. and Asher D.J., 2001. *MNRAS,* **327**, 126.

Jewitt, D.C., 2005. *Astron. J.,* **129**, 530.

Jewitt, D.C. and Fernandez, Y., 2001. *In:* M. Ya. Marov and H. Rickman, eds. *Collisional Processes in the Solar System.* Dordrecht: Kluwer, 143.

Jewitt, D.C., Chizmadia, L., Grimm, R. and Prialnik, D., 2007. *In*: V.B. Reipurth, D. Jewitt and K. Keil, eds. *Protostars and Planets.* Tuscon: Univ. of Arizona Press, 863.

Johnson, F.M., 1971. *Ann. New York Acad. Sci.,* **194**, 3.

Jones, B.W., Underwood, D.R. and Sleep, P.N., 2005. *ApJ,* **622**, 1091.

Junge, K., Eicken, H. and Deming, J.W., 2004. *Appl. Environ. Microbiol.*, **70**, 550.

Karl, D.M., Bird, D.F., Björkman, K. *et al.*, 1999. *Science,* **286**, 2144.

Kelvin, Lord (William Thompson), 1871. Report of the Forty-First Meeting of the British Association for the Advancement of Science, August 1871 Edinburgh, lxxxiv.

Kissel, J. and Krueger, F.R., 1987, *Nature*, **326**, 760.

Koyama, H. and Inutsuka, S.-I., 2000. *ApJ*, **532**, 980.

Krishna Swamy, K.S., 2005. *Dust in the Universe.* Singapore: World Scientific Publishing Co.

Krueger, F.R. and Kissel, J., 2000. *Stern und Weltraum*, **5**, 330.

Krueger, F.R., Werther, W., Kissel, J. *et al.*, 2004. *Rapid Comm. Mass Spectros.,* **18**, 103.

Kührt, E.K. and Keller, H.U., 1996. *Earth, Moon, and Planets*, **72**, 79.

Kuiper, G.P., 1951. *Astrophysics*. New York: McGraw-Hill, 357.

Kwok, S., 2009, *Astrophys. Sp. Sci.,* **319**, 5.

Lambert, L.H., Cox, T., Mitchell, K. *et al.*, 1998. *Int. J. Syst. Bact.,* **48**, 511.

Landgraf, M., Liou, J.-C., Zook, H.A. *et al.*, 2002. *Astron. J.*, **123**, 2857.

Leitch, E.M. and Vasisht, G., 1998. *New Astron.,* **3,** 51.

Leroux, H., Jacob, D., Stodolna, J. *et al.*, 2008. *American Mineralogist,* **93**, 1933.

Levin, G.V., 2007. *Int. J. Astrobiol.*, **6**(2), 95.

Levison, H.F., Morbidelli, A., Dones, L. *et al.*, 2002. *Science*, **296**, 2212.

Lewis, N.F., 1971. *J. Gen. Microbiol.,* **66**, 29.

Lin, L.-H., Wang, P.-L., Rumble, D. *et al.*, 2006. *Science*, **314**, 479.

Lindahl, T., 1993. *Nature,* **362**, 709.

Liou, J.-C., Zook, H.A. and Jackson, A.A., 1999. *Icarus*, **141**, 13.

Lisse, C.M., van Cleve, J., Adams, A.C. *et al.*, 2006. *Science,* **313**, 635.

Lodders, K., Osborne, R., 1999. *Space Sci. Rev.*, **90**, 289.

Love, S.G. and Brownlee, D.E., 1993. *Science*, **262**, 550.

Lyttleton, R.A., 1948. *MNRAS*, **108**, 465.

MacPherson, G.J., Davis, A.M. and Zinner, E.K., 1995. *Meteoritics*, **30**, 365.

Malfait, K., Waelkens, C., Waters, L.B.F.M *et al.*, 1998. *Astron. Astrophys.*, **332**, L25.

Manning, C.E., Mojzsis, S.J. and Harrison, T.M., 2006. *Am. J. Sci.,* **306**, 303.

Marcy, G.W. and Butler, R.P., 1998. *Ann. Rev. Astron. Astrophys.*, **36**, 57.

Marcy, G.W. and Butler, R.P., 1996. *Astrophys. J.,* **464**, L147.

Matese, J.J., Whitman, P.G., Innanen, K.A. and Valtonen, M.J., 1995. *Icarus,* **116,** 255.

Mathis, J.S., 1996. *Astrophys. J.*, **472**, 643.

Mattila, K., 1979. *Astron. & Astrophys.*, **78**, 253.
Mayor, M. and Queloz, D., 1995. *Nature*, **378**, 355.
McCaughrean, M.J. and O'Dell, C.R., 1996. *Astron. J.*, **111**, 1977.
McCrea, W.H., 1975. *Nature*, **255**, 607.
McIntosh, B.A. and Hajduk, A., 1983. *MNRAS*, **205,** 931.
McKay, D.S., Gibson Jr., E. K., Thomas-Keprta, K.L. *et al.*, 1996. *Science,* **273**, 924.
McSween, H.Y., 1979. *Geochim Cosmochim. Acta,* **43**, 1761.
Melnick, G., Neufeld, D.A., Saavik Ford, K.E. *et al.*, 2001. *Nature*, **412**, 160.
Melosh, H.J., 2003. *Astrobiology*, **3**, 207.
Melosh, H.J., 1989. *Impact Cratering: A Geologic Process.* New York: Oxford University Press.
Melosh, H.J., 1988. *Nature,* **332**, 687.
Merk, R. and Prialnik, D., 2003. *Earth, Moon and Planets,* **92**, 359.
Merrill, K.M., Russell, R.W. and Soifer, B.T., 1976. *Astrophys. J.*, **207**, 763.
Mie, G., 1908. *Ann. Physik*, **25**, 377.
Mileikowsky, C., Cucinotta, F.A., Wilson, J.W. *et al.*, 2000. *Icarus,* **145**, 391.
Miller, S.L., 1953, *Science,* **117**, 528.
Miller, S.L. and Urey, H.C., 1959. *Science*, **130**, 245.
Mojzsis, S.J., Arrhenius, G., McKeegan, K.D. *et al.*, 1996. *Nature*, **384**, 55.
Mojzsis, S.J., Harrison, T.M. and Pidgeon, R.T., 2001. *Nature*, **409**, 178.
Mopper, K., Kilham, K. and Degans, E.T., 1973. *Marine Biol.*, **19**, 323.
Morbidelli, A., Jedicke, R., Bottke, W.F. *et al.*, 2002. *Icarus,* **158,** 329.
Mostefaoui, S., Lugmair, G.W., Hoppe, P. and Goresy, A. El., 2004. *New Astronomoy Reviews*, **48**, 155.
Motta, V., Mediavilla, E., Muñoz, J.A. *et al.*, 2002. *ApJ,* **574**, 719.
Mumma, M.J., DiSanti, M.A., Russo, N.D. *et al.*, 1996. *Science*, **272**, 1310.

Nandy, K., 1964. *Publ. Roy. Obs. Edin*, **3**, 142.
Napier, W.M., 1987. *In*: Z. Ceplecha and P. Pecina, eds. *Interplanetary Matter: Proc. 10th European Regional Meeting in Astronomy*, Prague, **2**, 13.
Napier, W.M., 2007. *Int. J. Astrobiol.*, **6**, 223.
Napier, W.M., 2004. *MNRAS,* **348**, 46.
Napier, W.M., 2006. *MNRAS*, **36**(3), 977.
Napier, W.M. and Asher, D.J., 2009. *Astron. Geophys.* **50**(1), 18.
Napier, W.M. and Clube, S.V.M., 1979. *Nature*, **282**, 455.
Napier, W.M. and Staniucha, M., 1982. *MNRAS*, **198**, 723.
Napier, W. M., Humphries, C. M., 1986. *MNRAS*, **221**, 105.
Napier, W. M., Wickramasinghe, J. T. and Wickramasinghe, N. C., 2007. *Int. J. Astrobiol.*, **6**, 321.
Narlikar, J.V.N., Lloyd, D., Wickramasinghe, N.C. *et al.*, 2003. *Astrophys. Sp. Sci.*, **285**(2), 555.
Nelson, R.M., Soderblom, L.A. and Hapke, B.W., 2004. *Icarus*, **167,** 37.

Neukum, G. and Ivanov, B.A., 1994. *In:* T. Gehrels, ed. *Hazards due to Comets and Asteroids.* Tucson: Univ. of Arizona.

Nicholson, W.L., Munakata, N., Horneck, G. *et al.*, 2000. *Biol. Rev. Mol.*, **64**, 548.

Nisbet, E.G. and Sleep, N.H., 2001. *Nature*, **409**, 1083.

Nurmi, P., Valtonen, M. and Zheng, J.Q., 2001. *MNRAS*, **327**, 1367.

O'Keefe, J.D. and Ahrens, T.J., 1993. *J. Geophys. Res.*, **98**, 17011.

Okuda, H., Shibai, N., Nakagawa, T. *et al.*, 1990. *Astrophys. J.*, **351**, 89.

Okuda, H., Shibai, N., Nakagawa, T. *et al.*, 1989. *In*: M. Morris, ed. *The Centre of the Galaxy: Proc. IAU Symp. No. 136*, July 25–29 1988 Los Angeles. Dordrecht: Kluwer, 281.

Olsen, G.J. and Woese, C.R., 1997. *Cell*, **89**, 991.

Oort J. H., 1950. *Bull. Astron. Inst. Neth.*, **11**, 91.

Oort, J.H. and van de Hulst, H.C., 1946. *B.A.N.*, **10**, 187 (No. 376).

Oparin, A.I., 1953. *The Origin of Life* (trans. S. Margulis). New York: Dover Publications.

Öpik, E.J., 1932. *Proc. Amer. Acad. Arts Sci.*, **67**, 169.

Öpik, E.J., 1973. *Astrophys. Sp. Sci.*, **21**, 307.

Overmann, H., and Gemerden, H., 2000. *FEMS Microbiol. Rev.*, **24**, 591.

Overmann, J., Cyoionka, H. and Pfennig, N., 1992. *Limnol. Oceanogr.*, **33** (1), 150.

Pendleton, Y.J. and Allamandola, L.J., 2002. *ApJ Suppl.*, **138**, 76.

Pendleton, Y.J., Sandford, S.A., Allamandola, L.J. *et al.*, 1994. *ApJ*, **437**, 683.

Perrin, J.-M., Darbon, S. and Sivan, J.-P., 1995. *Astron. and Astrophys.*, **304**, L21.

Pflug, H.D., 1984. *In:* N.C. Wickramasinghe, ed. *Fundamental Studies and the Future of Science*, Cardiff: Univ. College Cardiff Press.

Raghaven, D., Henry, T.J., Mason, B.D. *et al.*, 2006. *ApJ*, **646**, 523.

Raup, D.M. and Sepkoski Jr., J.J., 1984. *Proc. Nat. Acad. Sci. USA,* **81**, 801.

Rickman, H., Fernandez, J.A., Tancredi, G. and Licandro, J., 2001. *In*: M.Ya. Marov and H. Rickman, eds. *Collisional Processes in the Solar System. Astrophys. Space Sci. Librabry Vol. 261*. Dordrecht: Kluwer, 131.

Rickman, H., Fouchard, M., Froeschlé, C. *et al.*, 2008. *Cel. Mech. and Dyn. Astron.* **102**, 111.

Rietmeijer, F.J.M., 2002. *In*: E. Murad and I.P. Williams, eds. *Meteors in the Earth's Atmosphere.* Cambridge: Cambridge Univ. Press, 215.

Rivkina, E.M., Friedmann, E.I., McKay, C.P. and Gilichinsky, D.A., 2000. *Environ. Microbiol,* **66**(8), 3230.

Roy, A.E., 1978. *Orbital Motion.* Bristol: Adam Hilger Ltd., 309.

Saavik Ford, K.E., Neufeld, D.A., 2001. *ApJ Lett.,* **557**, L113.

Salpeter, E.E. and Wickramasinghe, N.C., 1969. *Nature*, **222**, 442.

Sandford, S.A., 2008. *In*: S. Kwok and S.A. Sandford, eds. *Organic Matter in Space: Proc. IAU Symp. No. 251*, 18–22 February 2008 Hong Kong. Cambridge: Cambridge Univ. Press, 299.

Sandford, S.A., Aléon, J., Alexander, C.M.O.'D. *et al.*, 2006. *Science*, **314**, 1720.

Sapar, A. and Kuusik, I., 1978. *Publ. Tartu Astr. Obs.*, **46**, 717.

Savage, B.D. and Sembach, K.R., 1996. *Ann Rev. Astr. Astrophys*, **34**, 279.

Schalen, C., 1939. *Uppsala Obs. Ann.*, 1, No.2.

Schleicher, D.G., 2008. *Astron. J.*, **136**, 2204.

Secker, J., Wesson, P.S. and Lepock, J.R., 1994. *Astrophys. Sp. Sci.* **219**, 1.

Sivan, J.-P. and Perrin, J.-M., 1993. *ApJ*, **404**, 258.

Smith, J.D.T., Draine, B.T., Dale, D.A., *et al.*, 2007. *ApJ*, **656**, 770.

Southworth, R. B. and Hawkins, G. S., 1963. *Smithsonian Contributions to Astrophysics*, **7**, 261.

Spitzer, L., 1978. Physical Processes in the Interstellar Medium. New York:Wiley & Sons.

Stark, A.A. and Brand, J., 1989. *ApJ*, **339**, 763.

Stebbins, J., Huffer, C.H. and Whitford, A.E., 1939. *ApJ*, **90**, 209.

Stecher, T.P., 1965. *Astrophys. J.*, **142**, 1683.

Steel, D.I., 1993. *MNRAS*, **264**, 813.

Stein, W.A. and Gillett, F.C., 1971. *Nature Phys. Sci.*, **233**, 72.

Stetter, K.O., Fiala, G., Huber, G., *et al.*, 1990. *FEMS Microbiol. Rev.*, **75**, 117.

Stoffler, D., Horneck, G., Ott, S. *et al.*, 2007. *Icarus*, **186**(2), 585.

Stothers, R.B., 1998. *MNRAS*, **300**, 1098.

Stothers, R.B., 2006. *MNRAS*, **365**, 178.

Stothers, R.B., 1985. *Nature*, **317**, 338.

Svensmark, H., 2007. *A&G*, **48**, 118.

Szomouru, A. and Guhathakurta, P., 1998. *Astrophys. J.*, **494**, L93.

Szostak, J.W., Bartel, D.P. and Luisi, P.L., 2001. *Nature*, **409**, 387.

Taft, E.A. and Phillipp, H.R., 1965. *Phys. Rev.*, **138**, A197.

Tegler, S.C. and Romanishin, W., 2000. *Nature*, **407**, 979.

Temple, R., 2007. *Int. J. Astrobiol.*, **6**, 169.

Tepfer, D. and Leach, S., 2006. *Astrophys. Sp. Sci.*, **306**, 69.

Tielens, A.G.G.M., Wooden, D.H., Allamandola, L.J., *et al.*, 1996. *ApJ*, **461**, 210.

Urey, H.C., 1952. *Proc. Nat. Acad. Sci. USA*, **38** (4), 351.

Valtonen, M., Nurmi, P., Zheng, J.-Q. *et al.*, 2009. *ApJ*, **690**, 210.

Valtonen, M.J., 1983. *Observatory*, **103**, 1.

van Flandern, T.C., 1978. *Icarus*, **36**, 51.

von Helmholtz, H., 1874. *In:* W. Thomson and P.G. Tait, eds. *Handbuch de Theortetische Physik*, **1**(2), Brauncschweig.

Vanysek, V. and Wickramasinghe, N.C., 1975. *Astrophys. Sp. Sci.*, **33**, L19.

Vernazza, P., Mothé-Diniz, T., Barucci, M. A. *et al.*, 2005. *Astron. Astrophys.*, **436**, 1113.

Vidal-Madjar, A., Lagrange-Henri, A.-M., Feldman, P.D. *et al.*, 1994. *Astron. Astrophys.*, **290**, 245.

Vreeland, R.H., Rosenzweig, W.D. and Powers, D., 2000. *Nature*, **407**, 897.

Wachtershauser, G., 1990. *Proc. Nat. Acad. Sci. USA,* **87**(1), 200.

Waelkens, C. and Waters, L.B.F.M., 1997. *In*: Y.J. Pendleton and A.G.G.M. Tielens, eds. *From Stardust to Planetesimals ASP Conference Series*, **122**, 67.

Wainwright, M., Wickramasinghe, N.C., Narlikar, J.V. and Rajaratnam, P., 2003. *FEMS Microbio.l Lett.*, **218**, 161.

Wallis M.K. and Al-Mufti S., 1996. *Earth, Moon, and Planets*, **72**, 91.

Wallis, M.K. and Wickramasinghe, N.C., 2004. *MNRAS*, **348**, 52.

Wallis, M.K. and Wickramasinghe, N.C., 1995. *Earth Planet Sci. Lett.*, **130**, 69.

Wallis, M.K., 1980. *Nature*, **284**, 431.

Weingartner, J.C. and Draine, B.T., 2001. *ApJ*, **553**, 581.

Weisberg, M.K. and Connolly, H.C., 2008. *Proc. 39th Lunar and Planetary Science Conference,* 10–14 March 2008 League City Texas, 1918.

Weiss, B.P., Kirschvink, J.L., Baudenbacher, F.J. *et al.*, 2000. *Science*, **290**, 791.

Weissman, P.R. and Lowry, S.C., 2001. *Bull. Am. Astron. Soc.*, **33**, 1094.

Westley, M.S., Baragiola, R.A., Johnson, R.E. *et al.*, 1995. *Nature*, **373**, 405.

Whipple, F. L., 1950. *Astrophys. J.*, **111**, 375.

Wickramasinghe, D.T. and Allen, D.A., 1986. *Nature*, **323**, 44.

Wickramasinghe, D.T., Hoyle, F., Wickramasinghe, N.C. and Al-Mufti, S., 1986. *Earth, Moon and Planets*, **36**, 295.

Wickramasinghe, J.T., 2007. *PhD thesis*. Cardiff University.

Wickramasinghe, N.C., 1993. *In:* A. Mampaso, M. Prieto and F. Sanchez, eds. *Infrared Astronomy,* Cambridge University Press, 303.

Wickramasinghe, N.C., 1967. *Interstellar Grains*. London: Chapman & Hall.

Wickramasinghe, N.C., 1973. *Light Scattering Functions for Small Particles*. New York: Wiley.

Wickramasinghe N.C., 1974. *Nature*, **252**, 462

Wickramasinghe, N.C. and Hoyle, F., 1998. *Astrophys. Sp. Sci.*, **259**, 205.

Wickramasinghe, N.C. and Hoyle, F., 1999. *Astrophys. Sp. Sci.*, **268**, 379.

Wickramasinghe, N.C. and Wickramasinghe, J.T., 2008. *Astrophys. Sp. Sci.*, **317**, 133.

Wickramasinghe, N.C., Hoyle, F. and Al-Jubory, T., 1989. *Astrophys. Sp. Sci.*, **158**, 135.

Wickramasinghe, N.C., Hoyle, F. and Al-Jubory, T., 1990. *Astrophys. Sp. Sci.*, **166**, 333.

Wickramasinghe, N.C., Hoyle, F. and Lloyd, D., 1996. *Astrophys. Sp. Sci.,* **240**, 161.

Wickramasinghe, N.C., Lloyd, D. and Wickramasinghe, J.T., 2002. *Proc. SPIE*, **4495**, 255.

Wickramasinghe, N.C., Wainwright, M., Narlikar, J.V. *et al.*, 2003. *Astrophys. Sp. Sci.*, **283,** 403.

Wickramasinghe, N.C., Wickramasinghe, A.N. and Hoyle, F., 1992. *Astrophys. Sp. Sci.*, **196**, 167.

Wickramasinghe, N.C., Wickramasinghe, D.T. and Hoyle, F., 2001. *Astrophys. Sp. Sci.*, **275**, 181.

Wickramasinghe, N.C., Wickramasinghe, J.T. and Mediavilla, E., 2005. *Astrophys. Sp. Sci.*, **298**, 452.

Wilner, D.J., D'Alessio, P., Calvet, N. *et al.*, 2005. *ApJ,* **626**(2), L109.

Woese, C., 1967. *The Genetic Code.* New York: Harper and Row.

Woese, C. and Fox, G., 1977. *Proc. Natl. Acad. Sci.USA,* **74**(11), 5088.

Woolf, N.J. and Ney, E.P., 1969. *ApJ Lett.*, **155**, L181.

Yabushita, S., 1993. *MNRAS*, **260**, 819.

Yabushita, S., 1989. *MNRAS*, **240**, 69.

Yin, Q., Jacobsen, S.B., Yamashita, K. *et al.*, 2002. *Nature*, **418**, 949.

Zappalá, V., Cellino, A., Gladman, B.J., Manley, S. *et al.*, 1998. *Icarus*, **134**, 176.

Ziegler, K. and Longstaffer, F.J., 2000. *Clay and Clay Minerals*, **48**, 474.

Zook, H.A. and Berg, O.E., 1975. *Planetary and Space Sci.*, **23**, 183.

Zubko, V., Dwek, E and Arendt, R.G., 2004. *Astrophys. J. Suppl.*, **152**, 211.

Index